Bee Optimism:

Translational Research Can Rescue Honeybees And Other Pollinators

  Jay D. Evans

Essays written by Jay D. Evans, United States Department of Agriculture

Preface by Marla Spivak, University of Minnesota

Published by the International Bee Research Association and Northern Bee Books

All rights reserved. First edition, First Printing © Jay D. Evans

No part of this publication may be reproduced or transmitted in any form by any means, electronic, or mechanical, including photocopying, reproduction in print or electronic newsletters, recording devices or any information storage and retrieval systems know known to be developed, without permission in writing from the publisher, except by a reviewer who wishes to quote brief passages in connection with a review written for inclusion in a newspaper, magazine, broadcast, internet or other electronic broadcast system.

Earlier versions of these essays first appeared in Bee Culture Magazine, Kim Flottum, editor (A.I. Root, Medina, Ohio).

ISBN 978-0-86098-290-6

BEE OPTIMISM:

TRANSLATIONAL RESEARCH CAN RESCUE HONEYBEES AND OTHER POLLINATORS

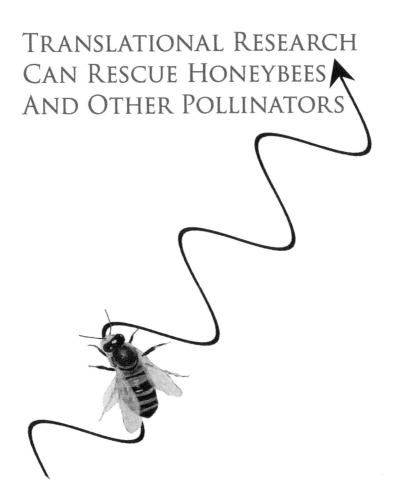

JAY D. EVANS

CONTENTS

INTRODUCTION

PART ONE
CHALLENGES AND OPPORTUNITIES

1. Colony Collapse Decade: Reflections on ten years of chasing a curse…..p. 3
2. Good bug, bad bug…..p. 7
3. Virus variety…..p. 10
4. Prime time…..p. 13
5. Let the Sonenshine in: New ways to control mites…..p. 16
6. Scraping out a living…..p. 19
7. Fat's domino effect…..p. 22
8. Genes, germs, and stress…..p. 26
9. Pros and cons of middle age…..p. 30
10. No country for old bees…..p. 34
11. Spring beauties and the beast …..p. 38

PART TWO
SWEETNESS AND LIGHT

12. Pollen counts: New ways to assess forage quality…..p. 42
13. Minerals and the bee's needs…..p. 46
14. Keeping your honey waiting…..p. 49
15. Resin d'etre: Can propolis be used by bees and beekeepers to improve colony health?…..p. 53
16. Next-generation scientists…..p. 57
17. Love in the time of chasmogamy: how and when do bees improve soybean yields?…..p. 62

PART THREE
ROYAL DECREES

18. Replacement queens, a true Cinderella story…..p. 67
19. History tends to repeat itself…..p. 71
20. Convergent ways to expose and fight mites…..p. 75
21. Holding the line on trait rot and inbreeding…..p. 79
22. Magic bullets for mites…..p. 83
23. Bugs with benefits…..p. 88
24. Island time and resistant bees…..p. 91
25. Bees, beenomes, and benefits from science…..p. 96
26. Social status and the single bee…..p. 99

PART FOUR
CLIMATE CHANGE

27. Weather, your bees live or die…..p. 103
28. Delayed mortification…..p. 107
29. Over in winter…..p. 112
30. Snowbirds, snow, and supplements shed light on overwintering success…..p. 116
31. Winter stirrings…..p. 120
32. Spring greenings…..p. 124

CLOSING

33. #Beeoptomism: What if the honey super really *is* half full?…..p. 129
34. #Beeoptomism 2.0: Let's not go viral!…..p. 133

Photo Credits…..138

ACKNOWLEDGEMENTS

Many thanks to Kim Flottum for the chance to write essays for *Bee Culture,* and for his encouragement through thick and thin. Also, to Kathy Summers for her editorial hand and to Jerry Hayes who has taken over the editorship at *Bee Culture* and who has long inspired many of us in the bee world. I also thank the International Bee Research Association and Jeremy Burbidge (Northern Bee Books) for their efforts and encouragement. I was inspired to start these essays by beekeepers of all scales and passions, from those keeping bees hidden behind their townhouses to the towering forces for commercial beekeeping whose work reflects the sweat of generations. For them, I tried to capture the work of the people I admire greatly, scientists trying to make practical sense of a complex world and those just trying to capture some of its beauty. I am always inspired by my colleagues at the USDA-ARS Beltsville Bee Research Laboratory, who push every day for advances that will help honey bees and the environment. We're still trying. Thanks to Dawn Lopez, Raymond Peterson, Philene Vu, Mervyn Eddie and many of the authors whose work I discuss, for reading through these essays. Special thanks to the talented Humberto Boncristiani (www.insidethehive.tv) for providing excellent photos and advice and to my family for giving meaning and the space to keep playing with bugs. Camille Getz drew the bee on the cover and section heads. These essays referenced current research articles at the time of writing. Should you wish to see the latest in those topics I would suggest using the article titles as a query in any web browser and checking new papers that cite these works and are available to you.

Dedicated to my mom, Camille, a constant teacher of biology, science, and goodness.

PREFACE

The title of this book reveals all: Jay Evans approaches science and life with clarity and optimism. Every essay in this book is concise yet funny, and rich in content. Each cleverly-worded chapter uncovers exciting findings from complicated research. Some topics covered in these 34 short essays include how bees' fat bodies may hold clues to new mite control strategies (the "Fats Domino Effect"), an aerial view of helicopter beekeepers and burned out foragers, why bees drink dirty water, chasmogamy in soybeans, "trait rot", genetic signatures described as chords composed of eight genetic notes, and winter bees as the third worker form. The essay "Magic Bullets for Mites" is a great example of Jay's ability to communicate complex topics with compelling, accessible prose.

Even though he muses that "we are attempting to manage a social organism that is affected by rules we do not yet fully understand" Jay remains firmly rooted in the #Beeoptimism movement: for his scientific mind, the unknown is full of possibilities. I thoroughly enjoyed this book and devoured it in one sitting, although I recommend savoring it more slowly.

        Marla Spivak
        St. Paul, Minnesota

INTRODUCTION

Practical research discoveries take a twisting path. Basic research driven by the human need to understand nature can, often years or even decades later, lead to huge advances that benefit people or the environment. Alternatively, sure-thing tests of a new product, management strategy, or breeding scheme often fail at the last moment when applied to 'real' life. In these essays, I describe success stories in translational bee research, projects where the toil of scientists has led to 'news you can use' as a beekeeper. These stories come from university, government, or industry scientists in the worldwide bee research community who have made an important and practical discovery, or who have put pieces together from others' research to make a substantial advance for bee science and beekeeping. In many cases these studies were carried out in direct partnership with beekeepers, or at the least with beekeepers as their inspiration. These are imperfect songs for science, which gives much more than it takes.

PART ONE

CHALLENGES AND OPPORTUNITIES

1

COLONY COLLAPSE DECADE: REFLECTIONS ON TEN YEARS OF CHASING A CURSE

We in the scientific world can bask in many battles won through the identification and treatment of honey bee challenges, and we do so. Still, it is hard for anyone to claim that we have won the war for bee health. I am just off the phone with another long-suffering commercial beekeeper who reports tremendous winter losses by him and his beekeeping colleagues, just as their bees are needed in California for their most important pollination event. This call comes more than ten years after he and others reported what became labeled Colony Collapse Disorder (CCD). Some of the forensic signs are the same; colonies with a laying queen but few foragers, or a forager force that declines sharply in a week or two, plenty of stored food, no mite issues, and a serious lack of dead bodies. In other cases, colonies simply don't build up, despite having all the eggs and brood needed to explode. The common thread seems to be a dearth of older bees. As a middle-aged researcher I feel a bond with middle-aged bees that drop dead just when they have a chance to bring treasure to their sisters. I am optimistic that beekeepers, scientists, and others who care for bees will win this war…but why does it have to be so hard?

As was the case in 2007, pests, pathogens, and chemical insults are the most likely suspects for current bee colony losses. But which of these, and why do they push some colonies and entire operations over the cliff and not others? Much has been written about synergies and compounded effects, and scientists are making progress tackling those synergies. However, we will never have the time or resources to march through and

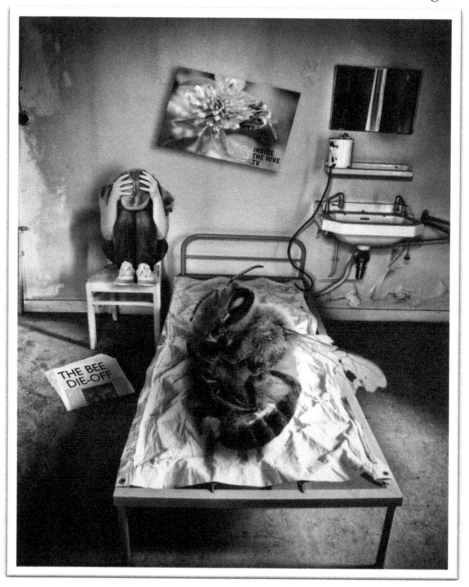

The despair of CCD, HFB

test each and every possible combination of living and non-living factors that impact honey bee health. This fact leads to two possible strategies. The first strategy is to design experimental tests after winnowing down the number of suspects and their toxic twins based on exposure, predicted interactions, or lucky forensic breaks. This strategy works and has been the basis for some insightful studies. For one, USDA Agricultural Research Service researcher Yu Cheng Zhu and colleagues measured the impacts of the common pesticide imidacloprid when alone or in conjunction with a set of seven compounds most likely to co-occur with imidacloprid in the environment (Zhu et al., 2017). The results helped predict which compounds are most damaging when delivered in concert. Zhu coupled these survival assays with enzymatic screens to determine which bee defenses were effective against each compound. Such studies are excellent for explaining the world and specific risks, and fall within the comfort zone of us scientists, but they are painstaking and therefore cannot address every biological and physical threat to honey bees. It has been argued that a more experiential, rather than experimental, approach is needed.

Farmers of all sorts tend to fall more into the experiential, or intuitive, fold. I was reminded of this by the recent death of my wonderful stepfather Melvin Getz, a long-term and very successful Colorado rancher. Mel had already retired from ranching when I met him twenty years ago but he had endless stories of the trade and they were fascinating. He could, "rope and throw and brand 'em," for sure. Still, along with an ability to understand and manage cows, he clearly succeeded through an innate ability to understand the external forces on cattle and their human keepers. Most of his working days were spent on non-livestock challenges including water and grazing rights, community connections, economics, and snowstorms. He and those in his extended family spent as much time studying these external issues and advocating for them as they did riding herd, and that was a requirement for staying successful. In my limited experience as an outsider, commercial beekeepers who have survived in the trade are the same way. They are super intuitive and aware that their livelihoods depend as much on their intuition for economics, outside challenges, and sociology as on their knowledge of bees. Dr. Sainath Suryanarayanan from the University of Wisconsin has put together an important set of writings describing how scientists and intuitive beekeepers think differently about the challenges at hand. Sai has studied both entomology and societal forces and he presents compelling arguments that both sides can be blinded by their biases, resulting in over-predicting (more often on the beekeeper side) and under-predicting (more often by those adhering to the standards of modern science) the degree to which the above stresses impact bees. One example he describes that struck home for me was the description by beekeepers and more intuitive scientists of 'spotty' brood patterns. A beekeeper with years of experience will know in their heart that something is wrong (and they are right) and will be tempted to associate this syndrome with something/anything out of the ordinary in their recent past. In contrast, many scientists will take this as an opportunity to reduce the fates of individual larvae to singular causes and might spend a lifetime, or at least a PhD, tackling a handful of potential causes. This doesn't necessarily mean that either side has bad intentions, just different ways of approaching a crisis. Sai, along with Daniel Kleinman, presents these and more arguments in a stimulating book, Vanishing Bees: Science, Politics, and Honey Bee Health (Suryanarayanan & Lee Kleinman, 2016). In the open-source spirit I will strive for in these essays, you can also immerse yourself in some of Sai's thinking for free, via YouTube at (https://www.youtube.com/watch?v=UlxCXa4DgjY). While I mostly reside in the comfort zone of established science and its culture, Sai and Mel, although they never met, have taught me there is another path toward our common goal of healthier bees. I do not promise that these essays will rival Shakespeare, but I think a quote from Hamlet fits nicely:

> "The time is out of joint.
> O cursed spite, That ever I was born to set it right!
> Nay, come, let's go together."

Zhu, Y.C., Yao, J., Adamczyk, J. & Luttrell, R. (2017). Synergistic toxicity and physiological impact of imidacloprid alone and binary mixtures with seven representative pesticides on honey bee (Apis mellifera). PLoS ONE, 12. 10.1371/journal.pone.0176837.

Suryanarayanan, S. & Lee Kleinman, D. (2016). Vanishing bees: Science, politics, and honeybee health.

"Be(e)ing Human: A Social History of Collapsing Beehives in the United States," YouTube video, 00:33:33, Posted by "Rachel Carson Center," Aug 8, 2013, https://www.youtube.com/watch?v=UlxCXa4DgjY.

2

GOOD BUG, BAD BUG

As I finish this essay on Super Bowl Sunday I am reminded of a sarcastic comment used to describe the game of soccer in the U.S., "Soccer is the game of the future…and it always will be." As a longtime soccer dad, I am ever hopeful that the world's game will reach its proper place at the top of local sports. I am not holding my breath and nor, apparently, is the rest of the country. The same might be said for probiotics. In human health and agriculture, probiotic treatments have received much attention over the years, leading to substantial research and development as well as commercial products. As a USDA employee, I cannot endorse or condemn the available commercial products, but I would like to discuss two research studies that provide new insights for the promising field of probiotics.

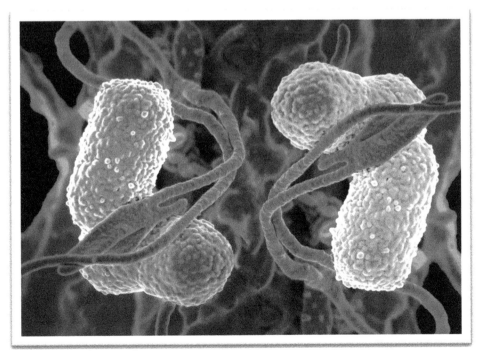

Microbes in their web, HFB

Both studies, one in Italy and one in the U.S., tested the impacts of bacteria found in the honey bee gut on levels of the honey bee parasite *Nosema ceranae*. In Italy, Loredana Baffoni and colleagues isolated and cultivated six lactic-acid bacteria found in honey bee guts (Baffoni et al., 2016). They then combined these into a bacterial cocktail that was hand-fed to newly emerged honey bee workers. These bees were contained in relatively sterile cages in the laboratory. Five days later, half of the bees were handfed an infective dose of *N. ceranae* spores in sugar water while half were given sugar water alone. Bees that had received the bacterial mix prior to being exposed to *Nosema* spores had significantly lower disease loads than did control bees raised on sugar water alone. When bees were not fed *Nosema* spores, those bees harboring natural *Nosema* infections also tended to have fewer spores when they had been primed with the bacterial cocktail, although this result was not statistically significant. Finally, in all sets there was substantial variation in *Nosema* spore loads that was not explained by the probiotic treatment. These promising results deserve field tests in an apiary.

Vanessa Corby-Harris and colleagues at the USDA's Carl Hayden Honey Bee Research Center in Tucson have described an additional candidate for a bacterium that has positive effects on bee health (Corby-Harris et al., 2016). *Parasaccharibacter apium* is found in both larval and adult bees. These researchers cultured an isolate of *P. apium* taken from bee larvae, and then set out to determine its impact on bee disease and bee health. Importantly, they first confirmed that this bacterium would remain viable when mixed into pollen patties under hive temperatures. Live bacteria survived well for at least 24 hours, arguably enough time to be consumed by bees in the hive. In addition, larvae from hives sustained with bacteria-rich pollen patties held more live bacteria than did larvae from hives with control pollen patties. They then tested the impacts of *P. apium* on *Nosema* disease. Newly emerged bees from hives fed *P. apium* patties and bees from control hives were again handfed *Nosema ceranae* spores. After these bees were maintained in sterile cages for ten days, bees from *P. apium* colonies showed lower *Nosema* loads than did bees from control colonies. While the results await field trials, *P. apium* shows great promise as a 'good' bug. A similar study carried out by Ryan Schwarz, now at Fort Lewis College in Colorado, provides a sobering reminder of the need to do follow-up work outside with actual be colonies. Working with Nancy Moran at the University of Texas and myself, Ryan showed a substantial decrease in levels of the gut parasite *Lotmaria passim* when bees were primed with the gut bacterium *Snodgrassella alvi*…at least when experiments were carried out in sterile cups. Unfortunately, just the opposite occurred when those same bees were marked and released into colonies (Schwarz et al., 2016). Inspired that it is at least possible to permanently tweak the gut microbes in bees through probiotics, Ryan and others are hot on the trail to find bacteria that have a positive impact in the field.

For those interested in a fresh take on the impacts, limits, and promise of gut microbes for *human* health (including when and how to eat a restorative poop sandwich), science writer Ed Yong has just written an entertaining book, "I Contain Multitudes: The Microbes Within Us and A Grander View of Life" (Yong, 2016). The similarities and challenges for human and agricultural probiotics are striking, from clear benefits in some cases to less exciting results elsewhere. I, for one, remain optimistic that a science-based approach to developing probiotics will lead to more products that improve honey bee health. Then again, I missed the end of what I hear was a decent Super Bowl after going to bed with dreams of soccer greatness.

Baffoni, L., Gaggìa, F., Alberoni, D., Cabbri, R., Nanetti, A., Biavati, B. et al. (2016). Effect of dietary supplementation of Bifidobacterium and Lactobacillus strains in Apis mellifera L. against Nosema ceranae. Beneficial Microbes, 7, 45-51. DOI: 10.3920/BM2015.0085.

Corby-Harris, V., Snyder, L., Meador, C.A.D., Naldo, R., Mott, B. & Anderson, K.E. (2016). Parasaccharibacter apium, gen. Nov., sp. Nov., Improves Honey Bee (Hymenoptera: Apidae) resistance to Nosema. Journal of Economic Entomology, 109, 537-543. https://doi.org/10.1093/jee/tow012.

Schwarz, R.S., Moran, N.A. & Evans, J.D. (2016). Early gut colonizers shape parasite susceptibility and microbiota composition in honey bee workers. Proceedings of the National Academy of Sciences of the United States of America, 113, 9345-9350. 10.1073/pnas.1606631113.

Yong, E. (2016). I Contain Multitudes: The Microbes Within Us and a Grander View of Life. Ecco, an imprint of HarperCollins Publishers, New York, NY.

3

VIRUS VARIETY

There is a time bomb ticking off the Brazilian coast. In the Open Access journal *Scientific Reports*, Laura Brettell and Stephen Martin present new data on virus levels and virus diversity in an extraordinary population of honey bees that has survived *Varroa* mites without treatment for decades (Brettell & Martin, 2017). First established in 1984, the honey bees of Isla Fernando de Noronha, far off the Atlantic coast of Brazil, are comprised of Italian honey bee stock from the United States (California queens inseminated by Georgia males). No bees were present on this extremely

Severe DWV infection, HFB

remote island, nor on its equatorial neighbors, and the population has yet to mix with Africanized bees or any other honey bee races.

The honey bees of Fernando de Noronha did not escape *Varroa* mites in 1984, and in fact, mite loads were extremely high early in the history of this population. David deJong and Espencer Soares measured average loads of 26 mites per 100 bees in 1991 and while those loads decreased in subsequent years, they remained far higher than sustainable mite loads in most honey bee populations (De Jong & Soares, 1997). As with the rest of Brazil, chemical controls were not used to control *Varroa* and these bees have flourished despite the presence of mites. In 2017, Brettell and Martin found that average mite loads from this population were substantially lower, up to 2 mites per 100 bees and in line with colonies in other parts of the world. No signs of viral disease or other symptoms of mite-related disease have ever been seen in these bees.

Brettell and Martin used sensitive genetic tests to screen colonies for Deformed wing virus (DWV). To their surprise, they found this mite-associated virus despite never seeing its eponymous trait in infected bees. DWV levels were extremely low, similar to levels seen in honey bee populations that have never been exposed to *Varroa* mites. Low virus levels and no ill effects could reflect virus resistance by the bees of Fernando de Noronha or exposure to viral strains that are not aggressively virulent. Brettell and Martin feel the latter scenario best explains the good health of these island bees.

Improved genetic tests have led to extensive surveys of diversity within DWV and the other major bee viruses. Most honey bee viruses persist as (appropriately) a swarm of variants, some of which are more damaging to their bee hosts, but all of which infect bees at some level. This variation reflects exceedingly high mutation rates for DWV and other viruses, followed by ongoing selection favoring winning viral types that can make the most copies in cells of their bee hosts. These successful viral strains go hand in hand with more damaging effects on bees. Viral winners are not universal at the level of continents, thanks to controls over the movement of bees, and instant successful DWV variants seem to arise spontaneously from within this viral group. There is even a more distant cousin, a highly successful DWV variant called DWV-B (or *Varroa destructor* virus), which is now causing widespread bee damage in much of Europe. As a further twist, Eugene Ryabov and colleagues have shown that recombinant viruses with a protein shell resembling DWV-B and inner enzymes closest to DWV-A (the worldwide 'classic' successful DWV strain) are the worst news for bees in England, growing to high levels and causing heavy disease symptoms (Moore et al., 2011). Viruses are asexual and it takes a unique and rare event for strains to merge together. Accordingly, when such mergers are seen they must reflect superior growth in bees, or a better ability to survive in bees or their mite vectors.

Back to the tiny island of Fernando de Noronha off the Atlantic coast of Brazil, Brettell and Martin showed a diverse assemblage of viruses in this population. Intriguingly, they did not see signs of any of the predominant viral strains found elsewhere in mite-infested bees. In fact, they saw what appear to be numerous mild viral strains at low levels, which seem tolerable to bee hosts. Their time bomb analogy comes from the fact that it might take only a freak mutation, or a human assist, to introduce a more virulent strain of DWV into this bee population. If that occurs, they suggest that 30-plus years of population stability in the face of *Varroa* could come to a sudden end. This last hypothesis, if true, could inform better ways to manage mites and viruses and, in particular, the regulation of bee and virus movement across populations.

Brettell, L.E. & Martin, S.J. (2017). Oldest Varroa tolerant honey bee population provides insight into the origins of the global decline of honey bees. Scientific Reports, 7. DOI: 10.1038/srep45953.

De Jong, D. & Soares, E. (1997). An Isolated Population of Italian Bees That Has Survived Varroa jacobsoni Infestation Without Treatment for over 12 Years. American Bee Journal, 137, 742-745.

Moore, J., et al. (2011). "Recombinants between Deformed wing virus and Varroa destructor virus-1 may prevail in Varroa destructor-infested honeybee colonies." Journal of General Virology 92: 156-161.

4

PRIME TIME

What if you could predict the disease environment of your offspring and then do something to increase their odds of surviving specific disease threats? This seems like a good deal, but the devil would be in the details…. How accurately could future threats be predicted? How costly would it be if things changed? Finally, how on earth could a mom predict the future, let alone send a warning to progeny they might never meet. Two recent studies suggest that queen bees can do just that, raising interesting research questions and suggesting novel ways that queen breeders might (might! these are early days) improve their product.

Honey bees and other insects receive little respect from medical and veterinary researchers studying immunity. It is accepted that insects possess an immune response but this response was for the most part seen as more like duct tape than like a Swiss army knife, something capable of recognizing and reducing bacteria and fungi but not in an especially sophisticated way. Begrudgingly, most scientists now accept that insect immunity is complex and capable of targeting specific pathogen groups with a range of strategies. This acceptance resulted from many excellent studies showing how insect immunity both improves insect health and reduces the risk to humans of parasites that are vectored by insects. Still, the research 'establishment' drew the line on the ability of insects to prepare for a future threat. While it is clear that the insect immune system does not have the same adaptive immune mechanisms found in our own bodies, it has been shown multiple times that insects can use prior experience with pathogens and parasites to be better prepared for later attacks. More controversial is whether that preparation crosses generations from parents to their offspring.

Paul Schmid-Hempel and his scientific progeny in Switzerland have been generating insights into bumble bee immunity for decades. They are driven by natural curiosity and perhaps a desire to prove wrong those who underestimate the insect immune system. Alongside lasting work showing how bumble bees and their parasites punch and counter-punch each other, they were the first to show that a queen's disease exposure early in life could impact how her worker bees survived their own disease threats. More recently, they have focused on the 'how' aspects of this discovery. Seth Barribeau, Schmid-Hempel, and Ben Sadd describe the immune outcomes of trans-generational immune priming in a recently published paper in the open-access journal *PLoS One* (Barribeau et al., 2016).

Prior to egg laying, one set of bumble bee queens was injected with a salt solution containing 2 million cells of a common soil bacterium, one set was injected with the salt solution alone, and one set was not injected at all. Adult worker bee progeny of these queens were further divided into those that were themselves injected with bacteria and those that received only a shot of salt. A key result from this study is that worker bees produced by queens exposed to bacteria showed high levels of known immune genes (the precursors of antimicrobial peptides). In fact, those bees showed a typical response to bacterial challenge, even when they themselves were not exposed to bacteria. This study also explored changes in these worker offspring more broadly, identifying many more similarities between the progeny of challenged queens and bees that had themselves been exposed to bacteria.

North American bumble bee, HFB

Bee Culture readers will likely be more interested to know whether transgenerational priming works for honey bee queens and their offspring. One recent paper provides evidence that this might be the case. Javier Hernandez-Lopez and colleagues in Austria challenged queens with heat-killed cells of *Paenibacillus larvae*, the cause of American foulbrood disease. For a study described in the *Proceedings of the Royal Society* (López et al., 2014), they injected queens with 2 million heat-killed *P. larvae* cells. They then raised larvae from these queens in the lab, feeding them an infective dose of *P. larvae*. Offspring produced by exposed queens survived substantially better in the face of *P. larvae* than did offspring from control queens. Further, larvae from the challenged queens showed higher numbers of a class of blood cells linked to immunity. While it appears certain that Mom is sending disease cues to her worker offspring, exactly how these cues are sent remains unclear.

So, can trans-generational immune priming be co-opted to improve honey bee health? It seems likely, although much work is needed in order to optimize this and make it an important tool. Should anyone try this approach they will have to be extremely patient, and I would **not** recommend using *P. larvae* unless you are absolutely sure you have killed all spores. Should these insights lead to improved bee immunity, this will be yet another applied breakthrough primed by work of a prior generation focused on uncovering fundamental facts of life.

Barribeau, S.M., Schmid-Hempel, P. & Sadd, B.M. (2016). Royal decree: Gene expression in trans-generationally immune primed bumblebee workers mimics a primary immune response. PLoS ONE, 11. http://journals.plos.org/plosone/article?id=10.1371/journal.pone.0159635.

López, J.H., Schuehly, W., Crailsheim, K. & Riessberger-Gallé, U. (2014). Trans-generational immune priming in honeybees. Proceedings of the Royal Society B: Biological Sciences, 281. http://rspb.royalsocietypublishing.org/content/281/1785/20140454.

5

LET THE SONENSHINE IN: NEW WAYS TO CONTROL MITES

Efforts to control *Varroa* mites are expanding at the USDA-ARS Bee Research Laboratory (BRL), thanks to a lucky break. Many students follow their Professors to new places when jobs change but it is rare for a Professor to tag along with their student. Professor Daniel Sonenshine's excellent PhD. student Noble Egekwu was recruited by the Bee Lab's Dr. Steven Cook to begin postdoctoral studies on the impacts of pesticides and mites on honey bees. While the paperwork unfolded for that move, Sonenshine had plenty of time to consider a move at the same time. Following his retirement from Old Dominion University in 2002 after 30+ years, Sonenshine maintained a very active research program on ticks as an emeritus professor there. He has, in fact, received more grants after retiring than while a full-time professor, and has continued to make great

Daniel Sonenshine, HFB

advances in tick biology and control. Tempted by the chance to make an impact on an entirely different problem, he let go of his University and active tick work to join the hunt for controls against bee mites. It did not hurt that Dr. Sonenshine had strong ties in Maryland and that he and his wife were open to a new venture closer to grandkids and family. As Sonenshine says, he did not plan this, but when doors open and you see new ways to be of service, you go through.

Prior to his recent arrival at the BRL, Sonenshine had never worked with *Varroa* mites although he does remember the name and story of these mites from his studies alongside tick and mite systematist George Wharton at Ohio State University. While he was late to the *Varroa* bandwagon, Sonenshine has had many noteworthy accomplishments studying the related ticks for over 50 years. For one, he collected the world's knowledge of ticks into a key two-volume textbook *The Biology of Ticks*, 1991 and 1993 and followed this with an expanded second addition co-edited with his friend and colleague Michael Roe (Sonenshine, 1991, 1993). Sonenshine's research on dog ticks, deer ticks and 'soft' ticks helped lay down many of the fundamentals of tick biology. He and his students described the key tick protein vitellogenin and then discovered the receptor for this protein, a 'guide' that helps the protein do its thing in the right places (eggs, for one). He then focused on the various ways amorous ticks meet up, leading to seven patents useful for controlling tick populations. In one success that only a scientist could brag about, he developed and attached a tick lure meant for the South end of cows headed North (Norval et al., 1996). This lure contains an attractive pheromone for male ticks mixed with a mite-killing chemical. Once attached (careful now!), this patch is successful in leading 95% of a cow's ticks to their death under the cow's tail. The location is key since it is one place not reachable by a self-grooming cow (unlike bees, cows rarely groom each other). More recently, he and his students designed and validated a 'TickBot,' a remote-controlled car designed to collect and kill ticks in nature (Gaff et al., 2015). After navigating with the help of an embedded wire for 60 minutes, the Tickbot effectively cleared mites from a public trail. Take that, Roomba! He continues to keep abreast of tick research and has helped lead efforts to sequence and understand the deer tick genome and to determine how these and other ticks recognize their animal hosts. In many ways, his breakthroughs in tick biology helped set the stage for some of the most promising new *Varroa* controls.

What does Sonenshine have in mind for *Varroa*? A true scientists' scientist, he arrives at the lab on most days of the week, dons his white lab coat and quietly gets to work. With Drs. Cook and Egekwu, along with University colleagues, he has been pursuing new ways to raise *Varroa* in the lab. This will enable better testing of control methods for mites, some of which are hard to work out in colonies. He has also focused on *Varroa* feeding behaviors, shedding light on how mites scrape and chew their bee hosts, feeding on and transmitting deadly viruses. He continues to build on his past insights into mite love-talk and is complementing the lab skills of Egekwu and Cook by identifying key chemicals used by passionate or hungry mites. The plan is to use this knowledge to trick mites and lure them to their deaths. His former student Noble has left to continue studies in the laboratory of Professor Jamie Ellis in Florida, but Dan has stayed behind this time to help keep the local mite efforts going. We should all be thankful for Sonenshine's longtime dedication to the art of science and for volunteering to make a difference.

Sonenshine, D.E. (1991, 1993). The Biology of Ticks. Oxford University Press.

Norval, R.A.I., Sonenshine, D.E., Allan, S.A. & Burridge, M.J. (1996). Efficacy of pheromone-acaricide-impregnated tail-tag decoys for controlling the bont tick, Amblyomma hebraeum (Acari: Ixodidae), on cattle in Zimbabwe. Experimental & applied acarology, 20, 31-46. https://link.springer.com/article/10.1007/BF00051475.

Gaff, H.D., White, A., Leas, K., Kelman, P., Squire, J.C., L.Livingston, D. et al. (2015). TickBot: A novel robotic device for controlling tick populations in the natural environment. Ticks and Tick-borne Diseases, 6, 146-151.https://doi.org/10.1016/j.ttbdis.2014.11.004.

6

SCRAPING OUT A LIVING

A great mentor of mine once described true scientific discoveries as ones where half of the people say, "that's impossible" while the other half say, "that's a fact the whole world knows." This contradiction reflects an innate trait of scientists. We are stubbornly skeptical and, when a seriously novel insight comes in to view, can be slow to embrace the messenger. This year's messenger of an important paradigm shift in bee health is Dr. Sammy Ramsey (even with his newly minted PhD., Sammy is steadfast in not being addressed as Sam or Samuel). Dr. Ramsey and colleagues at the University of Maryland, including his adviser Dennis vanEngelsdorp, teamed up with USDA and University of Florida scientists to study what, exactly, *Varroa* mites take from their honeybee hosts. With a set of clever and tedious experiments they, to use a scientific term, "knocked it out of the park" for this topic. If you haven't heard the punch-line for this study, you can find the full paper online (Ramsey et al., 2019). It turns out *Varroa* mites are scrapers and not slurpers, carvers and not bloodsuckers, more

Varroa mite on worker bee, HFB

Jeffrey Dahmer than Dracula. This great insight brings past observations into better focus and sets the stage for breakthroughs in both fundamental and applied bee biology.

So, what did they learn and how? The study began by observations of where mites feed. It has long been known that older female mites initiate feeding on bee brood by carving a sort of artesian well from the outer skeleton of their hosts, returning to the same spot and eventually bringing their hungry offspring. This is distinct from, say, mosquitoes and aphids, which poke and prod their hosts repeatedly and in a way that reflects their fear of being whacked by their animal hosts or being eaten by a predator, respectively. Why does *Varroa* devote so much effort to one spot? Similarly, mites feeding on adult bees take time to find a protected niche between hardened plates and truly nurture their feeding spot, scraping away the hard shell and then settling in for what was long thought to be a drinking binge on bee hemolymph (watery blood). Ramsey and colleagues noticed that mites were selecting crevices that not only protected them, but gave an entry to the most accessible sources of bee fat. Strikingly, mites were most likely to be found directly over an ample reserve of the 'fat body,' a honey bee organ that is more complex and important than its name. This diffuse organ not only stores fat to be used for future efforts, but also contains cells that are key for detoxification, immunity, and general bee development. If you've tried to drink a thick milkshake through a thin straw you will know that tapping into the body's fat reserve is not the fastest route for draining blood. In contrast, it is an excellent spot for carving out a complete and filling meal. Varroa mouthparts are entirely consistent with this feeding strategy, looking like little rasps and rakes, with no evidence of straws of any sort. Ramsey and colleagues present amazing microscopic images of Varroa mites and their bee hosts. Along with close-ups of mite mouthparts, they were able to show precisely where mites had scraped their entry into bee bodies. They even show pieces of fatty food left behind by feeding mites whose meals (and lives) ended under a million-dollar microscope.

The researchers had me convinced right there, but they stacked their paper with additional clever and convincing arguments. As a shout-out to pre-teenage boys everywhere, they also discuss mite excreta. Everyone poops, but the solid nature of *Varroa* feces, as opposed to the honeydew of aphids, is further evidence of a diet rich in fat. They nailed this observation by feeding adult bees with both a water-soluble dye and a dye that attaches to bee fat. As expected, mites collected after a feeding bout were full of the fat-linked dye and showed very little signs that they had even touched the liquid parts of their hosts. If mite poop is your thing, Francisco Posada-Florez and USDA colleagues recently provided an in-depth analysis of this byproduct (Posada-Florez et al., 2018). Even at the closest look available for poop science, all evidence supports the feeding model proposed in Ramsey's paper.

If mites are feeding on bee fats, is this really good for them, or are they just dealing with an easily accessible and chewable part of the bee? Using diets ranging from pure bee 'blood' to pure bee fat, Ramsey and colleagues found a distinctly better outcome with the latter. In fact, mites fed on anything less than 50% fat body (the consistency of chowder) never survived a seven-day trial, while those fed 100% fat body (too many fast-food analogies to name) did the best of all. These mites survived better AND laid more eggs, indicating that, in nature, the most successful mites are those that tap directly into this organ. This result is immediately relevant for scientists who maintain *Varroa* mites in the laboratory, as they search for weaknesses and new treatments along with new insights into the biology of these odd creatures.

The next few years will tell if a better understanding of *Varroa* feeding habits will speed the search for new control methods. This line of thought should also provide new insights into how *Varroa* mites and associated viruses cause so much harm to bees. My guess is that fat removal and fat body damage is key to understanding why parasitized bees live shorter and less productive lives. In the next essay, I will explore how these insights into mite feeding help explain the devastating effects of unsolicited fat removal on honey bees. What is certain is that hard-won discoveries like this one remind many of us of how we found our thrill for science in the first place.

Ramsey, S.D., Ochoa, R., Bauchan, G., Gulbronson, C., Mowery, J.D., Cohen, A. et al. (2019). Varroa destructor feeds primarily on honey bee fat body tissue and not hemolymph. Proceedings of the National Academy of Sciences of the United States of America, 116, 1792-1801.

Posada-Florez, F., Sonenshine, D.E., Egekwu, N.I., Rice, C., Lupitskyy, R. & Cook, S.C. (2018). Insights into the metabolism and behaviour of Varroa destructor mites from analysis of their waste excretions. Parasitology, 1-6.

7

FAT'S DOMINO EFFECT

I have highlighted the insightful work by Dr. Samuel Ramsey showing that *Varroa* mites slice their way into and devour the fat bodies of honey bees (Varroa destructor feeds primarily on honey bee fat body tissue and not hemolymph, (Ramsey et al., 2019)). "Fat body" does not do these tissues justice. While they are a gathering place for lipids and proteins thought to give honey bee resources for the future, the fat body turns out to be a complex tissue, and arguably organ, that is essential for bee health.

In many ways, the fat body is like the human liver. For one, it drives the process to generate and process fats and other energy reserves, a key metabolic trait of our own livers. Like the liver, the fat body also makes proteins that help protect the body from pesticides and even pathogens like viruses and bacteria. In other ways, the insect fat body is quite unique. In larvae and pupae, the fat body is spread across much of the insect body from head to tail. In adults, the fat body is centered in the abdomen, but is still very much a cloud of cells over a large (for a bee) area. While it plays a similar role, the insect fat body is more like liver paté than the liver you would picture from a chicken or mammal.

So, what does it mean to make and break fat? Lipid droplets, or "fat pills" to use a less technical term, are components of fat body cells that act like small machines for both collecting and breaking down fat (or lipids). These fat machines respond to signals from the bee's body that indicate whether times are 1) favorable for storing energy or 2) a bit tighter, in which case stored fats have to be metabolized, or "burned", by enzymes to provide needed energy. In a sense there is no single on-switch for this whole process, the entire fat body is reading the lay of the land inside a bee, with each lipid droplet advancing or retreating in terms of energy release depending on local currents. As if that's not weird enough, the cumulative effects of these very local decisions are not limited to a bee's get-up-and-go. In fact, the switches going on in fat body cells can make or break some of the key traits of bee social behavior, from reproduction to the provisioning of nestmates with food to the switch from nurse to forager. For example, Vanessa Corby-Harris and colleagues have shown that the breakdown of fats in fat body cells can explain why nurse bees lose the very organs that make them good nurses. Nurse bees produce the jelly fed to developing larvae using specialized glands in their heads. Over

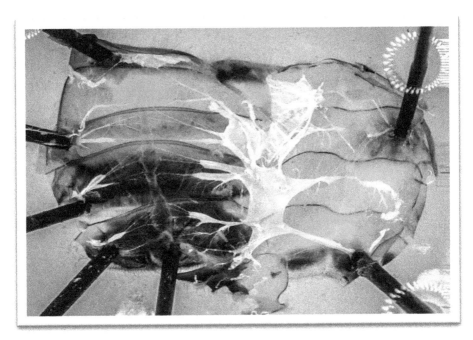

Dissected bee showing fat cell (whitish area), HFB

several studies, including "Fat body lipolysis connects poor nutrition to hypopharyngeal gland degradation in *Apis mellifera*," *(Corby-Harris et al., 2019)*, Corby-Harris and her team build the case that the decay of these glands, long known to occur when worker bees are under food stress, is driven in part by signals from fat cells on the other side of the bee's body.

Fat body cells also play a complicated role in bee disease. The cells themselves harbor viruses known to be damaging to bee health, including Deformed wing virus. These viruses reproduce by feasting on the resources found in fat body cells. At the same time, fat body cells release many of the protein components of bee immunity, attacking viruses and other pathogens in all parts of the bee body. Even more remarkably, fat body cells themselves might migrate to deal with threats to bee integrity. In a provocative recent paper focused on the much-studied fruit fly Drosophila, ("Fat body cells are motile and actively migrate to wounds to drive repair and prevent infection," (Franz et al., 2018)), Anna Franz and colleagues show that fat body cells actually perceive and then 'swim' to distant wounds in their own body. These are not rapid responses. In some cases, cells took hours to move towards a nearby wound. Still, the very fact that cells both move and, in the authors' words, 'multitask' by clogging wounds and releasing immune proteins, adds to the lore of the fat body. Overall, recent research suggests that many disease battles are won or lost within the squishy band of fat body cells.

Now that we know Varroa mites specifically target fat body cells in their quest for nutrients, there is an increasing sense that honey bee fat bodies are a key focal point for bee diseases and mortality. Some time ago, Gennaro di Prisco ("A mutualistic symbiosis between a parasitic mite and a pathogenic virus undermines honey bee immunity and health," (Di Prisco et al., 2016)) and colleagues proposed that the Varroa x virus x bee triangle is complicated by both mites and viruses ganging up on their bee hosts. In short, mites appear to benefit from immune-beating efforts of the viruses they carry. In their model, viruses released by feeding mites suppress a key ability of bees to seal up wounds, in effect increasing the success of feeding mites by keeping the feeding site open. It will be interesting to see how this model evolves now that we know mites are directly collecting fat body cells, rather than the blood surrounding these cells. Maybe some clever scientist will show that the slow race of fat body cells to a wound site is even slower when those cells are burdened by viruses. Or, perhaps, mite-damaged or devoured fat cells will break some sort of circuit in the fat body, limiting the strength of these cells to work together to regulate bee energy, immunity, or survival in the face of chemicals. Fat bodies have been understudied compared to other organs, perhaps because they look boring on the surface. Now that the secret of their importance is out, fat bodies should receive more scrutiny. They are bee sentinels, ready to send energy where needed one day and then race to the scene of a damaging injury the next, all while being attacked by ravenous mites and sneaky viruses.

Ramsey, S.D., Ochoa, R., Bauchan, G., Gulbronson, C., Mowery, J.D., Cohen, A. et al. (2019). Varroa destructor feeds primarily on honey bee fat body tissue and not hemolymph. Proceedings of the National Academy of Sciences of the United States of America, 116, 1792-1801.

Corby-Harris, V., Snyder, L. & Meador, C. (2019). Fat body lipolysis connects poor nutrition to hypopharyngeal gland degradation in Apis mellifera. Journal of Insect Physiology, 116, 1-9.

Franz, A., Wood, W. & Martin, P. (2018). Fat Body Cells Are Motile and Actively Migrate to Wounds to Drive Repair and Prevent Infection. Developmental Cell, 44, 460-479.

Di Prisco, G., Annoscia, D., Margiotta, M., Ferrara, R., Varricchio, P., Zanni, V. et al. (2016). A mutualistic symbiosis undermining bee health. Proceedings of the National Academy of Sciences of the United States of America, 113, 3203-3208.

8

GENES, GERMS, AND STRESS

Honey bees are a resilient beast at the colony level, able to survive wholesale losses of worker bees to predators, weather, disorientation and often simply employee burnout after a few weeks of foraging. When most parts of the system are in order, small colony events that take away workers, brood, or food stores are reliably absorbed and the colony persists. These insults are not without some cost, though, and several recent studies tease apart how stress and disease acting on individual bees can lead to disfunction at the colony level.

First, consider disease. Infected bees can suffer physical damage ranging from weakened stomach walls in the case of *Nosema* infection to non-functional wings in the case of certain virus infections. Infected bees also pay an energetic cost when fighting off disease with an immune response, and might even suffer collateral damage on their own bodies through this response. Célia Bordier and colleagues in France have tackled the long-term effects on worker bees following an immune response ("Stress decreases pollen foraging performance in honeybees," (Bordier et al., 2018). By using a non-pathogenic challenge, i.e., simply damaging the hard, outer skin of bees, they were able to isolate the 'stress' cost of fighting disease from the actual cost of an infecting parasite. The results show a long-term change in behavior that is relevant to colony health. Most importantly, immune-challenged bees were half as likely to return from foraging trips with pollen. It is unclear whether they returned with nectar in place of pollen, or failed completely and came home with an empty honey stomach. When immune-challenged bees DID return with pollen, it took them 30% longer, suggesting they had issues with flight or navigation. The authors argue that changes in foraging success on this level, and shifts in what was brought back, can have long-term effects on the optimization of colony stores. While bees are known to compensate for gaps in hive resources (e.g., the ratios of stored pollen versus nectar), adding a bias against pollen foraging into the mix might compromise their investment decisions.

Another major stress for all pollinators comes from the quality of available forage. Adam Dolezal and colleagues explored the costs of poor flower resources on the abilities of bees to fend off viruses, a hot topic for bees in nature as well as managed bees, "Interacting stressors matter: diet quality and virus infection in honeybee health," (Dolezal et al., 2019a). In their trials using bees caged in incubators, bees fed one type of pollen, from the Mediterranean plant *Cistus*, survived viral infection better than bees fed no pollen, but substantially worse than bees fed chestnut pollen or a blend of several more pollens. As in prior work, pollen supplements reduced virus levels overall when compared to bees fed simply carbohydrates. At the colony level, those with induced virus infections lost bees at a higher rate and losses were twice as severe in virus-infected colonies given no pollen versus those given any of the pollen supplements. Just this month, Belén Branchiccela, along with my USDA colleague Miguel Corona and others, showed similar impacts of pollen type on bee health in a field experiment involving commercial apiculture in Uruguay. This open-access paper, "Impact of nutritional stress on the honeybee colony health," (Branchiccela et al., 2019), describes a season of migratory beekeeping, during which half of the colonies were boosted

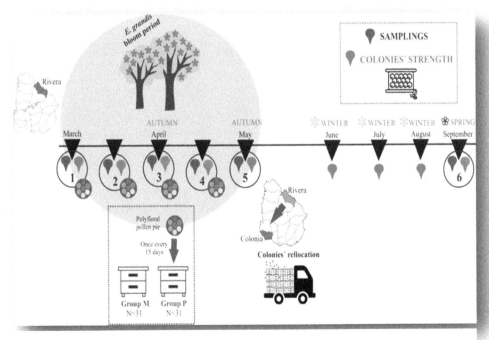

Migratory Beekeeping, BB

with pollen patties containing polyfloral pollen (graphic, this essay). Despite ample natural forage, supplemented colonies showed increased brood production and colony sizes along with lower *Nosema* loads. As a neat observation, colonies NOT given the polyfloral protein boost increased their own collection of diverse pollens, perhaps indicating a colony awareness of the need for these pollens.

So, what is new on the impacts of chemical stress on colony behavior? Théotime Colin and colleagues measured the effects of imidacloprid exposure on colony behaviors and traits in the article "Traces of a neonicotinoid induce precocious foraging and reduce foraging performance in honey bees," (Colin et al., 2019b). Larvae exposed to a standard low-dose exposure to this chemical (5 parts per billion) commenced foraging sooner than normal and finished their lives with substantially fewer foraging trips, a shift that these authors have argued can greatly reduce the long-term prospects of bee colonies. Interestingly, in a follow-up paper just this month, "Long-term dynamics of honey bee colonies following exposure to chemical stress," (Colin et al., 2019a), the authors found inconsistent impacts of thymol and imidacloprid stress on colonies, in fact showing an increase in colony size for bees exposed to imidacloprid in one site. These results could reflect the ability of honey bee colonies, some of the time, to respond to insults and come back with an adaptive group behavior. All of these studies, by expanding lab results into colony level dynamics, were challenging and the results were often not predicted. Still, given the beautiful beast that is a honey bee colony of thousands of individuals, it is essential to see how the beast, as a whole, responds to injuries suffered by its members.

Bordier, C., Klein, S., Le Conte, Y., Barron, A.B. & Alaux, C. (2018). Stress decreases pollen foraging performance in honeybees. Journal of Experimental Biology, 221.

Dolezal, A.G., Carrillo-Tripp, J., Judd, T.M., Allen Miller, W., Bonning, B.C. & Toth, A.L. (2019). Interacting stressors matter: Diet quality and virus infection in honeybee health. Royal Society Open Science, 6.

Branchiccela, B., Castelli, L., Corona, M., Díaz-Cetti, S., Invernizzi, C., Martínez de la Escalera, G. et al. (2019). Impact of nutritional stress on the honeybee colony health. Sci Rep., 12.

Colin, T., Meikle, W.G., Wu, X. & Barron, A.B. (2019). Traces of a Neonicotinoid Induce Precocious Foraging and Reduce Foraging Performance in Honey Bees. Environ. Sci. Technol, 53, 8252-8261.

Colin, T., Meikle, W.G., Paten, A.M. & Barron, A.B. (2019). Long-term dynamics of honey bee colonies following exposure to chemical stress. Science of the Total Environment, 677, 660-670.

9

PROS AND CONS OF MIDDLE AGE

It is well known that honey bee workers work inside the hive for the first week of their lives, with occasional tentative flights for those which are more precocious. By the end of their second week as adults, the healthy ones will start flying on a regular basis and, with luck, they will begin to bring pollen, nectar, resins, and water home against formidable odds. Thanks to two recent studies that have pushed the technology for 'helicopter beekeeping,' we now know quite a bit more about how bees perform as foragers.

Simon Klein, with colleagues on three continents, has described a fascinating study of the foraging careers of hundreds of worker honey bees. While this study can't really answer the key question of how bees die, the authors provide odds of death that would make insurance agents green with envy. They also describe the food incomes for both typical and atypical (or 'elite') foragers. The study is freely available at *Scientific Reports* under the title, "Honey bees increase their foraging performance and frequency of pollen trips through experience," (Klein et al., 2019). As the title implies, the first message is that foragers seem to get better at what they do as they age, good news for those of us who plan to work until we are found on our backs in the grass. They discovered this by using one of the most complex 'hive-monitoring' schemes yet. First, thousands of newly emerged bees in each of two colonies were geared up with tiny Radio-frequency identification (RFID) tags on their backs. These tags weigh only 1 mg, or $1/2000^{th}$ of a U.S. dime (1% of a typical worker bee's weight, superglue included). They are unique to each bee and trip an RFID antenna every time that particular bee passes by. The investigators then set up two one-way tunnels for the bees, one leaving the hive and one returning, by channeling bees with tubes equipped with opposing plastic bristles. Finally, these in- and out- express lanes were equipped with high resolution video cameras and balances to allow a visual inspection for pollen loads and a weight measurement to determine the success and failure of each trip. Fully geared up, the bees were set loose in Australian pastures.

The first surprise to me was how few foraging trips the average bee makes during her lifetime. Bees from the two colonies used in this study clocked 17 and 21 trips on average, respectively, and only rarely kept at it for more than a week. In fact, the average effective forager lifespan from these trials from first true foraging trip (> 10 seconds in duration) to lights out was under 5 days. During these trips, only one quarter of bees returned with pollen at least once, and all bees that were observed returning with pollen had at least one additional trip where they returned with just nectar, water, or no food at all. It was challenging to determine precisely how much food was harvested per trip, since departing bees both burned energy and likely defecated while on their journeys.

On the topic of old age and experience triumphing over youthful energy, Klein and colleagues found that bees steadily increased their rate of successful trips up until day 9 of foraging, if they were lucky enough to live that long. They also carried back larger loads, and more frequently returned with pollen. Why is this, when surely the energetic costs and various environmental hazards of foraging surely must accumulate? Are

Foraging bee, HFB

older bees more 'fit' perhaps because of prior journeys across the landscape? Or are they simply better able to locate and efficiently gather up food from flowers and make it home efficiently? Bees are known to refine their abilities to sniff out flowers over their foraging lifespans, and this perhaps best explains the higher success rates of older foragers. Abby Finkelstein and colleagues in Arizona and Germany provide a beautiful exposé of this fact in another open-access paper, "Foraging experiences durably modulate honey bees' sucrose responsiveness and antennal lobe biogenic amine levels," (Finkelstein et al., 2019). This study relied on a key test used to assess the learned and innate honey bee responses to signals in their environment, the 'Proboscis Extension Response.' This test is the honey bee equivalent of Pavlovian drooling, whereby individual bees are constrained and then exposed to a variety of cues ranging from tactile (perhaps sugar solutions brushed against one of their antennae) to airborne (any number of smells, from bomb residues to queen pheromones and floral scents). Exposed bees ably perceive and learn from these cues and, when the cues are linked to a food reward, respond by uncurling their proboscis 'tongues' to feel for that food.

Finkelstein and gang did not measure foraging directly, but instead established split cages of worker bees where all bees were in contact via gaps in a mesh screen but half of the bees were allowed to 'forage' on internal feeders while the other half received food only by mouth-to-mouth feeding from those foraging bees, through the mesh screen. They added a bit of aversive conditions for some of the foraging bees (a mild electric shock). This is a standard tool for determining just how influential a learned cue is on bee behavior, since even bees are hesitant to do something when pain is part of the reward. They then divided food (sucrose solution) sources into those with a floral scent and those without, and used these same scents as the triggering cue for bees asked to extend their proboscis tongues. Both foraging and confined bees showed an increased proboscis response to sugar scented with floral cues, but this response was substantially stronger for the forager bees, where scented nectar gave a threefold higher response rate than unscented nectar. This recall of floral scents was even stronger over time, with 'forager' bees that were tested several days later showing a higher response rate to scented foods. As evidence that these bees really had 'tuned' their senses to the best available cues, bees trained on a floral scent were actually less responsive to straight sugar than were bees which only knew pure sugar from the start. These learned biases were not evident in bees exposed to mild shocks, nor in bees which only received food from their nestmates, despite the fact that nestmate-fed bees presumably tasted the floral scents when they received food from the honey stomachs of their sisters.

Back to the great outdoors, could the higher rate of successful trips reflect learned cues from specific patches of flowers? Bees certainly use scents to key in on patches of flowers (if they didn't, flowers would not spend so much energy making these scents), but can bees become specialists over time, making a bee-line straight to familiar scents? Maybe. One other fascinating tidbit from the Klein study was the affirmation of 'elite' bees, which collected disproportionately high resources, both in terms of their number of trips and yields of nectar and pollen per trip. Building on this, it will be fascinating to see how well elite bees were targeting scents or individual patches that had been winners in the past. Curious minds will also want to know if these elite bees represent specific genetics (patrilines) or perhaps received lucky breaks in terms of their own nutrition or avoidance of abundant chemical or disease stresses. These studies and more, including rampant helicopter beekeeping by hobbyists and commercial beekeepers using a range of publicly available spying devices, will continue to shed light on what defines a good life for a worker bee.

Klein, S., Pasquaretta, C., He, X.J., Perry, C., Søvik, E., Devaud, J.-M. et al. (2019). Honey bees increase their foraging performance and frequency of pollen trips through experience. Scientific Reports, 9.

Finkelstein, A.B., Brent, C.S., Giurfa, M. & Amdam, G.V. (2019). Foraging Experiences Durably Modulate Honey Bees' Sucrose Responsiveness and Antennal Lobe Biogenic Amine Levels. Scientific Reports, 9.

10

NO COUNTRY FOR OLD BEES

Last essay I reviewed fundamental studies that have used hand-tagged bees to determine with actuarial precision the typical worker bee lifespan and productivity. These studies rely on Radio-frequency identification (RFID), a tagging system that beats other methods by using extremely light paper tags. RFID tags, unique to each bee, respond to an antenna signal every time tagged bees pass a hive recorder. An author of one of these RFID studies, Andrew Barron from Australia's Macquarie University (http://andrewbarron.org/), pointed me to a brand-new study from their group. This study combined the power of RFID with careful colony-level field experiments to determine the subtler effects of chemical stress on the endurance of foraging workers. Already oddly obsessed with the forces that lead to worker burnout, I was hooked. It turns out that this is one of many studies that have used RFID to spy on both overachieving workers and those that don't bring home the pollen. These studies are pushing the limits for our understanding of the subtler causes of honey bee colony failure.

Worker bee receiving an individual RFID tag, AB

Student Théotime Colin led the Macquarie group with guidance from Barron along with USDA (William Meikle) and Jiangxi University (Xiaobo Wu) collaborators. Their paper, "Traces of a Neonicotinoid Induce Precocious Foraging and Reduce Foraging Performance in Honey Bees," (Colin et al., 2019b), details the impacts of a typical chemical stress on lifetime food collection by worker bees. The punchline is that the effects are subtle but cumulative for colony health. What is truly fascinating, though, is the detail with which each worker's life is mapped out. As in Barron's earlier worker, 'elite' foragers carried out hundreds of trips, even in the colonies facing a chemical challenge. The proportion of these elites in the forager pool was lower when colonies were given 5 parts-per-billion imidacloprid in their food. Consequently, the net number of foraging trips (money in the form of honey and pollen) was significantly reduced. Exposed bees returned from 46 foraging trips on average, versus 64 trips for controls. Since they also foraged for shorter timespans, exposed bees spent nearly six hours less time on foraging trips over their lifespans than did controls (910 minutes versus 1437 minutes). The main reason for this was that exposed bees died in the midst of foraging, with a median foraging career of 8 days versus 10 for controls.

Shortened lives are also found in worker bees infected with disease. Lori Lach and colleagues showed that relatively low-level infections with the gut parasite Nosema (in their case 400 spores of *Nosema apis*) led to higher mortality and smaller returns to the colony. They present their work in the Journal of Invertebrate Pathology (Lach et al., 2015). Again, the effects were significant but subtle. The odds of taking even one foraging trip were lower in bees given the parasite, and challenged bees had significantly fewer trips across their lifetimes. They also much preferred nectar to pollen on foraging trips, while control bees showed an equal preference for pollen and nectar. This might be a rare case where foragers put their own needs ahead of the colony, since Nosema presents an energetic drain to foragers that can be countered by sugar consumption.

Virus infections, even those with ominous names like Deformed wing virus and Acute bee paralysis virus (or 'Slow' bee paralysis virus, for that matter), are generally not visible to beekeepers. Nevertheless, like Nosema and chemical stress, viral infections can have strong impacts on individual

RFID-tagged bee on flower, AB

bees and colonies. A recent paper by Kristof Benaets and colleagues in Belgium and England tackles the long-term impacts of Deformed wing virus on worker bees (Benaets et al., 2017). Perhaps not surprisingly by now, infected bees showed both a tendency to die young and poor foraging returns when they survived. Tagged bees that had been injected with a sublethal virus dose were twice as likely to die before they reached the age of foraging, despite starting to forage at a younger age. As foragers they fared even worse, flying three fewer days than healthy bees. In total, infected bees lived nearly five days less than controls, and that lost time was especially noticeable in their abilities to bring home food.

As alarming as a shortened lifespan might be, nutrition, stress, and disease can also change the abilities of bees to perform critical benchmark roles as they develop. These changes might throw the whole colony out of balance. Barron, again, described this adeptly in an essay titled "Death of the Bee Hive" (Barron, 2015). Normally, an active forager cohort in bee hives acts as a break for younger bees, arresting their development so that they remain as nurses and hive bees for up to two weeks. When foragers die young, the breaks are released and workers begin to forage early. They are not so good in their new role and that fact, combined with a tendency to burn out early as foragers, can cause colonies to spiral downward.

Colin, T., et al. (2019). "Traces of a Neonicotinoid Induce Precocious Foraging and Reduce Foraging Performance in Honey Bees." Environ. Sci. Technol 53(14): 8252-8261.

Lach, L., et al. (2015). "Parasitized honey bees are less likely to forage and carry less pollen." Journal of Invertebrate Pathology 130: 64-71.

Benaets, K., et al. (2017). "Covert deformed wing virus infections have long-term deleterious effects on honeybee foraging and survival." Proceedings of the Royal Society B: Biological Sciences 284(1848).

Barron, A. B. (2015). "Death of the bee hive: Understanding the failure of an insect society." Current Opinion in Insect Science 10: 45-50.

11

SPRING BEAUTIES AND THE BEAST

This was to be an article on May flowers, another expression of my (well hidden, even to me) insights into the plants that are everything to bees and our food chain. Indeed, early garden seeds are in the ground as I write this, trees are attracting bees, and you can taste and smell spring. When this book comes out in May, the salad greens we've planted will be ready to gulp down, and I can't wait. Nevertheless, I am unable to think of much beyond infectious diseases right now, and viruses in particular. COVID-19 has upended many of our lives and has already brought its share of tragedy to the world. COVID-19 also comes with a worldwide cloud of worry, as we fear infections ourselves, sharing infections with neighbors and loved ones, and the effects this is having on our remarkable species as a whole.

One of the saddest parts of the current pandemic is the imperative for social isolation. I don't mean the "spring breakers" who are curtailed from Florida beaches. They might well benefit from less sunburn and less indulgence in liquids that are best used to sterilize their hands. I am thinking of our normal circles: contacts with friends and family, caring for neighbors, chats on the street, hugs. My own young-adult daughter is self-isolating at someone else's house, making me long for the days of looking at her eating cereal in the morning, completely undefended against an attack of dad jokes. Thank goodness for the grid of electronic communication. If that falters by the time you read this, just know that the silencing of those jokes will hit hard (the latest was the guy trying to telework saying, "My coworker has just LOST it, she chases the mailman, pees on the neighbor's property, and won't come back inside, I'm calling HR!").

What is truly missed worldwide is the ability to stop by and check on elderly neighbors, the abilities of spouses and friends to visit people in assisted living, rehab, or prison settings, and the million other ways we as a species help each other. All of these are more challenging in the short term, although brave caretakers are doubling down to make sure everyone is emotionally and physically attended to. It is also fine to mourn the silencing of bee clubs, wherein the most senior of the club's members would right now (March) be showing newbies how to install packages, and would in May be arguing about splits, how early they can possibly treat for mites, and how to get a few pounds of honey from those first packages. These meetings will be back, and for now there are many thoughtful translations of this wisdom. I like "Beekeeping: Your First Years" which my Bee Culture editors seem to think I will need indefinitely, and they are right!.

Honey bee on flower in pollinator garden, PG

Enough about our species, this is a bee magazine and there are plenty of smarter people discussing the human predicament. Instead, I am thinking of hopeful signs that honey bees can and do solve their own disease challenges. As beekeepers know, honey bee colonies take the bad with the good. They tower over other pollinators not just through individual smarts (although they have plenty of that, as leading bee researcher Dr. Gene Robinson has shown, https://lab.igb.illinois.edu/robinson/) but through their relentless cooperation. While most bee species thrive on solitude, a solitary honey bee would lack purpose in all ways. Bees in a healthy colony can direct thousands of nestmates to flowers or nest sites and can defend against bears that outweigh them by 2,000,000 times (yes,

I looked that up). At the same time, honey bees live in close quarters and share microbes in ways that make germophobes cringe. Entire bee colonies can be brought down by disease, and they are, but that is not the norm. Despite being attacked by everything from viruses to bears, honey bees persist. They have even worked out the social isolation thing. As a group, hygienic bees enforce social isolation. This group of bees, appropriately 'nurses' for the most part, protect the mother ship just like our immune cells do within us, and our nurses and doctors do at the group level. At some risk to themselves, they remove diseased bees, groom parasites and spores from their nestmates, and generally fight back against the biological enemies of the hive. How they do this has been a topic of study since the days of Walter Rothenbuhler and students at Ohio State University, and the subtleties of nurse bee diagnostics and care continue to be revealed. Hygienic behavior depends on bee diagnostics. Not the types that employee bee scientists and commercial outfits, but the kind bees do on their own. Like other animals, honey bees pick up smells from their environment and the perception of these smells primes them to act. If you want a quick review of bee pheromones check out an excellent Youtube lecture on "Applying The Basics Of Honey Bee Biology" by my esteemed colleague Dr. Clarence Collison, from minute "15:00" onward. This lecture and several linked ones on bee biology by Dr. Collison are well worth the time, even when you aren't practicing social distancing.

Honey bee on flower in pollinator garden, PG

Sujin Yi and colleagues at Incheon University, including longtime bee researcher Hyung Wook Kwon, recently updated the world on cues used by bees to sniff out American foulbrood disease(Lee et al., 2020). Three volatile (smelly) compounds released by sick larvae trigger reactions in nurse bees and presumably help them tend to these larvae while ignoring the many healthy ones. My searching of the internet also shows that two inspiring researchers, Drs. Kaira Wagoner and Olav Rueppell from the University of North Carolina-Greensboro (Dr. Rueppell is now at the University of Alberta), have recent patented, "Methods and compositions for inducing hygienic behavior in honey bees." Given these advances there will likely be additional practical outcomes from the decades of studying how bees react to and suppress disease threats from within. That is good news for breeding and medicating, a hopeful sign for a challenging spring.

I am not sure there is a lesson in this for us, as we face our own pandemic, but I have hope. First, our own 'noses' in terms of disease diagnostics are enviable, and the knowledge derived from our ability to identify and act is power. Second, I would put a lot of trust in, and thanks for, our nurses and doctors, and anyone else who is honestly working to calm and help us in this or any crisis. Finally, there is nothing like a crisis to remind us we are all in this together, and our actions, good and bad, impact humanity. There has been an overwhelming amount of shared science during this crisis and that has to prime some solutions. Be well bee friends, spring is still spring and beekeepers old and new will get through this. Help a nestmate if you can.

The Robinson Lab, Honey Bee Research at the University of Illinois. Received at https://lab.igb.illinois.edu/robinson.

Applying The Basics Of Honey Bee Biology by Dr. Clarence Collison, Received at https://www.youtube.com/watch?v=pULjCfKX8_k.

Lee S, Lim S, Choi Y-S, Lee M-l & Kwon HW (2020) Volatile disease markers of American foulbrood-infected larvae in Apis mellifera. Journal of Insect Physiology 122: 104040. doi:https://doi.org/10.1016/j.jinsphys.2020.104040.

Methods and compositions for inducing hygienic behavior in honey bees. Received at https://patents.google.com/patent/US10524455B2/en.

PART TWO

SWEETNESS AND LIGHT

12

POLLEN COUNTS: NEW WAYS TO ASSESS FORAGE QUALITY

A clear understanding of the range and diversity of pollen collected by honey bees is important for improving colony health. Pollen from different species differs in protein content and other nutritional traits, and in some cases a multi-floral blend of pollen might give bees a stronger start than ample levels of any single species. Scrutinizing bee-collected pollen also provides an efficient way to survey available flowering plants for science and policy decisions. In fact, were it not for the bees, such surveys would require many researcher-lifetimes spent watching flowers and adding up pollen sources…and it would still require some guesswork to determine the realized value of these pollen sources for bee foragers.

Palynology, the visual study of pollens in order to assign species, is a longstanding field in bee research and plant biology generally. Trained palynologists can discriminate between hundreds of pollen types, and this method has had a big impact on assigning current and past plant distributions. As one example, the Discover Life consortium and Professor Debbie Delaney at University of Delaware have published a helpful visual guide for identifying some North American pollens (http://www.discoverlife.org/mp/20q?guide=Pollen). Gene-based strategies provide a complimentary and extremely effective approach for identifying the plant sources of pollen collected in the environment. Variable DNA regions or 'barcodes' can precisely link pollen grains to the plants that produced them. What has card-carrying palynologists worried for their livelihoods is the fact that even hugely complex pollen mixtures can be screened relatively cheaply and accurately using DNA. These genetic techniques are now showing their worth in practical studies of the food sources collected by honey bees and other pollinators.

On the technical side, a gene-based survey of pollen faces two major challenges, the equal extraction of DNA from pollens of all species and the presence of a genetic region that distinguishes all species in a given environment. For the latter, there are strong cases currently for several genetic regions that can be used as barcodes to identify pollen sources. I know my limits and won't try to rank these, but reputable bee researchers have shown the worth of using the Ribulose bisphosphate carboxylase

Forager with pollen, HFB

large chain gene (rbcL; Karen Bell and colleagues, (Bell et al., 2017)) and the ribosomal RNA internal transcribed spacer region (ITS, Robert Cornman and colleagues, (Cornman et al., 2015)). Rodney Richardson and colleagues have contrasted several genetic regions against microscopic counts in an attempt to show their strengths and weaknesses (Richardson et al., 2015). Similarly, Smart and colleagues (Smart et al., 2017) pitted informative genetic markers against microscopic analyses by knowledgeable palynologists. These studies show some clear challenges involving DNA extraction and 'equal representation' faced when using gene-based techniques versus microscopy. Nevertheless, both studies highlight the vast and quantitative insights gained from a well-planned genetic screen.

So, which questions are being addressed using a genetic approach to count pollen? One timely question involves honey bee foraging in the highly productive 'Prairie pothole' region of eastern North and South Dakota. This area is a home base for many of the leading commercial beekeepers, in large part thanks to an abundance of natural forage. 'Wildflowers' in this region have led to productive honey flows for decades and remain a key resource for the industry. Work led by federal agencies, including the U.S. Geological Survey (USGS) and US Department of Agriculture (USDA), along with the University of Minnesota, has focused on maintaining and improving this resource. Some of this work involves mapping changing land-use patterns on a regional scale (e.g., Clint Otto and colleagues, (Otto et al., 2016)). Land-use decisions involve a complex analysis of many factors, including agricultural practices, weather patterns, and the presence of managed and wild bees. Important insights for these decisions are coming equally from field observations as well as computer-based and genetic studies. Gene-based counts of pollen collected by bees, including the recent work in the Dakotas by Smart and colleagues, are playing an increasing role in making sound beekeeping and land management decisions. Beekeepers in the future might use pollen ID's along with colony survival and honey yields to determine when and where to place their hives. There is some poetry there, in that this is yet another case of harnessing the immense honey bee workforce to give humans a helping hand.

Discover Life| All Living Things| Pollen. Retrieved from http://www.discoverlife.org/mp/20q?guide=Pollen.

Bell, K.L., Loeffler, V.M. & Broshi, B.J. (2017). An rbcL Reference Library to Aid in the Identification of Plant Species Mixtures by DNA Metabarcoding. Applications in Plant Sciences, 5. doi:10.3732/apps.1600110.

Cornman, R.S., Otto, C.R.V., Iwanowicz, D. & Pettis, J.S. (2015). Taxonomic characterization of honey bee (Apis mellifera) pollen foraging based on non-overlapping paired-end sequencing of nuclear ribosomal loci. PLoS ONE, 10. doi:10.1371/journal.pone.0145365.

Richardson, R.T., Lin, C.H., Quijia, J.O., Riusech, N.S., Goodell, K. & Johnson, R.M. (2015). Rank-Based Characterization of Pollen Assemblages Collected by Honey Bees Using a Multi-Locus Metabarcoding Approach. Applications in Plant Sciences, 3. doi: 10.3732/apps.1500043.

Smart, M.D., Cornman, R.S., Iwanowicz, D.D., McDermott-Kubeczko, M., Pettis, J.S., Spivak, M.S. et al. (2017). A comparison of honey bee-collected pollen from working agricultural lands using light microscopy and its metabarcoding. Environmental Entomology, 46, 38-49. doi: 10.1093/ee/nvw159.

United States Geological Survey Pollinator Library. Retrieved from https://www.npwrc.usgs.gov/pollinator/home.

Otto, C.R.V., Roth, C.L., Carlson, B.L. & Smart, M.D. (2016). Land-use change reduces habitat suitability for supporting managed honey bee colonies in the Northern Great Plains. Proceedings of the National Academy of Sciences of the United States of America, 113, 10430-10435. doi: 10.1073/pnas.1603481113.

13

MINERALS AND THE BEE'S NEEDS

Rachael Bonoan and colleagues at Tufts University have been studying the impacts of salt and other minerals on honey bee health for several years. This work is spurred by historical observations that honey bees often choose murky water sources over pristine ones, arguably reflecting a search for resources that are scarce in pollen and nectar. Writing in *Ecological Entomology* (Bonoan et al., 2017), the authors describe a clever bee 'cafeteria' used to build the case that bees on water-foraging trips can identify minerals and selectively drink those waters highest in specific minerals, from salts to calcium. Their results hint at a seasonal shift in forager preferences for specific minerals coinciding with changes in pollen availability and consumption. While more could be done to determine whether bees actively search out specific minerals or are simply less choosy at certain times of the year, the results do suggest that water-collecting bees, like those foraging for pollen or nectar, are somehow driven by the hive mind.

Pierre Lau and James Nieh in San Diego took a lab-based approach to address similar questions (Lau & Nieh, 2016). Their study relied on a behavioral technique widely used for everything from age-based learning to measuring the impacts of pesticides on bees. They used this technique, the Proboscis Extension Reflex (PER), to determine the concentrations at which worker bees favor or avoid minerals found in water. They, too, showed a general preference for salty water, at concentrations far higher than levels observed in nearby water sources. Interestingly, individual bees from the same hive differed greatly in their desire for salt. In statistical terms, the salt preferences among bees from the same hive were far more variable than were average differences across hives. This result suggests that current nutritional status for those bees, past foraging experiences, or even genetics, can impact the desire for salt and other minerals. These authors also found that minerals both attracted and repelled worker bees. As an applied output for this result, the authors suggest that bees could be warned off from toxic waters if these waters were also marked with high levels of repellent minerals such as potassium.

Both studies provided a happy opportunity to revisit the elegant and practical nutrition research of Dr. Elton Herbert, Jr., a USDA-Beltsville honey bee scientist in the 1970's and early 1980's who died too young. Herbert was conversant in all components of the honey bee diet and used a series of experiments to test the favorability of minerals for bee health. In developing what would become the 'Beltsville bee diet,' Herbert and Hachiro Shimanuki showed that a mid-range level of minerals in an artificial bee diet led to greater growth than did low- or high-salt mixes (Herbert & Shimanuki, 1978). Herbert also tested the impacts of pollen-based versus 'mined' minerals on bee health (Herbert, 1978), showing that pollen ash derived from a custom blend of pollens led to the strongest brood growth when compared to a low-mineral diet or one supplemented with Wesson's salt, a product with >10X as much sodium. In contrast to the recent studies, bees were LESS likely to consume mineral-rich diets versus control diets lacking those minerals.

Supplementing a plant-based diet with alternative sources of minerals is of course not limited to bees. Butterflies and other insects routinely collect minerals from puddles and other impure water sources. Similarly, entire industries have been built around efforts to attract and/or sustain salt- or mineral-crazed domestic and free-living mammals. In an

Proboscis Extension Reflex (PER) assay, HFB

analogous 'cafeteria' study, Jose Estevez and colleagues in Spain showed that deer change their preferences for several minerals across seasons, reflecting their needs (Estevez et al., 2010). Similarly, Pablo Gambín and colleagues showed that red deer engage in 'osteophagia', or the consumption of antlers and bones, during seasons when calcium needs are highest (Gambin et al., 2017). As a further extreme, deer and other large grazers occasional harvest *living* animals, feeding on their skeletons when the available forage does not provide enough calcium. As with the honey bees, observations of mammalian feeding behavior have led clever researchers to new insights into dietary needs and new recommendations for increased growth. Hopefully this work will further improve the many artificial diets and supplements available for honey bees.

Bonoan, R.E., Tai, T.M., Tagle Rodriguez, M., Feller, L., Daddario, S.R., Czaja, R.A. et al. (2017). Seasonality of salt foraging in honey bees (Apis mellifera). Ecological Entomology, 42, 195-201. http://onlinelibrary.wiley.com/doi/10.1111/een.12375/pdf.

Lau, P.W. & Nieh, J.C. (2016). Salt preferences of honey bee water foragers. Journal of Experimental Biology, 219, 790-796. doi:10.1242/jeb.132019.

Herbert, E.W. & Shimanuki, H. (1978). Mineral Requirements for Brood-Rearing in Honeybees Fed a Synthetic Diet. Journal of Apicultural Research, 17, 118-122. http://dx.doi.org/10.1080/00218839.1978.11099916.

Herbert, E.W. (1978). A New Ash Mixture for Honeybees Maintained on a Synthetic Diet. Journal of Apicultural Research, 18, 144-147. http://dx.doi.org/10.1080/00218839.1979.11099958.

Estevez, J.A., Landete-Castillejos, T., Garcia, A.J., Ceacero, F., Martinez, A., Gaspar-Lopez, E. et al. (2010). Seasonal variations in plant mineral content and free-choice minerals consumed by deer. Animal Production Science, 50, 177-185. DOI: 10.1071/AN09012.

Gambin, P., Ceacero, F., Garcia, A.J., Landete-Castillejos, T. & Gallego, L. (2017). Patterns of antler consumption reveal osteophagia as a natural mineral resource in key periods for red deer (Cervus elaphus). European Journal of Wildlife Research, 63. DOI 10.1007/s10344-017-1095-4.

14

KEEPING YOUR HONEY WAITING

As a scientist I spend most of my time focused on a narrow slice of research devoted to honey bee health. Still, it is fun to take a break and read articles on topics way outside my comfort zone. So, it was with an upcoming article in *Food Chemistry* by Ioanni Pasias and colleagues in Greece, "Effect of late harvest and floral origin on honey antibacterial properties and quality parameters" (Pasias et al., 2018). To be honest, this title jogged memories of pulling frames from a neglected dead out, extracting honey and, much later, musing that the honey tasted a bit funky. As a hobbyist, somewhat funky honey is acceptable to my (non-paying) customers. In fact, friends will never tell you to your face that your honey tastes funky, even as they purse their lips and agonize over gentler words to describe it. Since I have a reputation to maintain, I draw the line at not poisoning anyone. The Pasias article promised to say, for one class of honey, whether anything dangerous might arise from such 'late harvest' honeys.

Pasias and his Argonauts measured the physicochemical traits of 38 honeys collected from a region of Greece. Their main focus was on two collections from the same beekeeper and apiary. One of these, 'Early Argos,' consisted of four honey samples extracted promptly at the end of the season. In contrast, honey for four 'Late Argos' samples remained in the comb in closed boxes for an additional year. All 38 honeys, along with three samples of Manuka honey from New Zealand, were subjected to a battery of tests, measuring everything from antimicrobial activity to levels of 5-Hydroxymethylfurfural (HMF) and diastase. HMF is a standard for estimates of honey aging and temperature spikes. In the end, HMF levels were substantially higher in the 'late' samples when compared to both fresh extracted samples from the same site (>12-fold higher) and 30 local honey samples purchased at markets (3x higher on average). Interestingly, the 'late' honey also showed greater antimicrobial properties than the early samples. In fact, the stored honey also showed two-fold higher antibiotic activity towards bacteria than did the famous Manuka honey. This is counterintuitive since it is generally believed that antimicrobial traits of honey diminish over time, even as honey itself can last many hundreds of years when undiluted. The Greek study had a weakness in that palynological analyses revealed the stored 'late' honey came from somewhat different floral sources than the 'early' sample. While the HMF levels and antimicrobial traits of this late honey were higher than any of the other polyfloral honeys form the area, it is not possible to say for certain how much of this difference was due to aging in the field versus a difference in sources...so onward to finding a better source.

As honey producers know, there is an extensive literature on the effects of heat and storage on honey traits. Dr. Clarence Collison wrote an excellent review on this in *Bee Culture*, adding an analysis of the various ways that sugar supplements can be detected in honey (Collison, 2016). The increase in HMF (the best-known adverse honey component) is not particularly fast for bottled honey, maybe an increase of 3 ppm per month on the high end. On the legal side, honeys subjected to lengthy storage of high heat, or lengthy storage, might run afoul of international standards for HMF and the loss of diastase activity (www.fao.org). Allowable levels for retail honey are under 40 ppm HMF and over 8 Schade units, diastase activity, with some exceptions. For one, these standards are relaxed to 80 ppm for honeys with naturally high HMF levels, including Manuka honey and tropical polyfloral honeys.

Honey from the Bee Research Lab collection, DL

The fastest route to high HMF levels in honey and in high fructose syrup fed to bees involves heat. Corn syrup stored at 40°C (104°F, admittedly quite warm) acquires HMF at a rate of 20 ppm per month while syrup subjected to temperatures of 49°C accumulates HMF at around 150 ppm/month (LeBlanc et al., 2009). Honey likely behaves in a similar way. Light exposure has a significant impact on honeys that are already bottled and out in the open. Our own honey collection is full of samples that are significantly darker today than when they were first collected several years ago showing the effects of a multi-year shelf life under fluorescent lighting. More studies are needed to determine adverse effects of light exposure on bottled honeys, although there is reason for concern since HMF is in fact a product of darkening in the form of the Maillard reaction.

You may be wondering if HMF or other degradation products in stored honey will impact your bees when honey frames are recycled back into hives. Zirbes and colleagues in Belgium have taken some of the guesswork out of that by reviewing levels at which HMF impacts bees directly (Zirbes et al., 2013). Based on this, a downside to moderately aged honey for bees seems unlikely. Assuming HMF accumulation occurs at a rate of 3 ppm/month, it would take three years to approach 100 ppm in honey, and the first signs of negative effects in bees tend to occur above that. In a very careful study involving worker bees, Blaise LeBlanc and colleagues showed a significant adverse impact of HMF in corn syrup at concentrations exceeding 250 ppm (LeBlanc et al., 2009), while Sophie Krainer and colleagues did not see adverse effects of HMF on bee larvae until levels in brood food were > 2000 ppm (Krainer et al., 2016). On the other hand, HMF could interact with the other stresses facing bees, including other honey traits impacted by age and temperature. Leslie Bailey, a great, and recently deceased, bee scientist, suggested this very thing over 50 years ago. He showed that honeys stored four and eight years had negative effects on caged bees fed this as their sole food source (Bailey, 1966). HMF levels in these honeys were not likely to be toxic, but other degradative processes seem to have had a cumulative effect that was bad for bees. Overall, storing honey in the dark at moderate temperatures for months, and even years, seems a good bet for bees as well as humans. On the other hand, if you neglect these boxes for several years, and if wax moths, mice, and beetles do not ravage them first, the resulting honey could be toxic for your bees.

Along with the usual U.S. government disclaimer that I cannot tell readers whose honey or syrup to buy, I want to make perfectly clear that I am NOT a food chemist, nor a food safety expert, so any insights related from these studies should be taken with caution with respect to food quality, palatability, or risk (eater beware!). Hopefully the cited studies will help those who want to explore the traits of aging honey in more detail. I, for one, will keep consuming, funky taste or not.

Pasias, I. N., et al. (2018). "Effect of late harvest and floral origin on honey antibacterial properties and quality parameters." Food Chemistry 242: 513-518.

Collison, C. (2016). A Closer Look: Feeding Sugar Syrup/HMI. Bee Culture. http://www.beeculture.com/a-closer-look-feeding-sugar-syruphmi/.

www.fao.org/input/download/standards/310/cxs_012e.pdf.

LeBlanc, B.W., Eggleston, G., Sammataro, D., Cornett, C., Dufault, R., Deeby, T. et al. (2009). Formation of hydroxymethylfurfural in domestic high-fructose corn syrup and its toxicity to the honey bee (Apis mellifera). Journal of Agricultural and Food Chemistry, 57, 7369-7376. https://doi.org/10.1021/jf9014526.

Zirbes, L., Nguyen, B.K., De Graaf, D.C., De Meulenaer, B., Reybroeck, W., Haubruge, E. et al. (2013). Hydroxymethylfurfural: A possible emergent cause of honey bee mortality? Journal of Agricultural and Food Chemistry, 61, 11865-11870. https://doi.org/10.1021/jf403280n.

Krainer, S., Brodschneider, R., Vollmann, J., Crailsheim, K. & Riessberger-Gallé, U. (2016). Effect of hydroxymethylfurfural (HMF) on mortality of artificially reared honey bee larvae (Apis mellifera carnica). Ecotoxicology, 25, 320-328. https://doi.org/10.1007/s10646-015-1590-x.

Bailey, L. (1966). The Effect of Acid-Hydrolysed Sucrose on Honeybees. Journal of Apicultural Research, 5, 127-136. https://doi.org/10.1080/00218839.1966.11100146.

15

RESIN D'ETRE: CAN PROPOLIS BE USED BY BEES AND BEEKEEPERS TO IMPROVE COLONY HEALTH?

Having visited Brother Adam's apiaries at Buckfast Abbey and his honey sites in lonely Dartmoor, I was struck by how orderly and thoughtful the setting was, a truly meditative place to breed a holistic bee. Brother Adam applied the science of the time and a healthy travel budget to do just that. One thing he thought bees could do without was propolis (plant resins laboriously collected by worker bees, stripped from their bodies, and applied to hive surfaces). Despite increased interest in the benefits of propolis for human health in the first half of the 1900's, the concept that collected plant resins could aid bees was lost on Brother Adam. Instead, he saw the downside for beekeepers dealing with hesitant removable frames and consequently sought breeding stock that left plant resins to the plants.

Top frames with propolis, HFB

Coincidentally, just before Brother Adam returned to Buckfast Abbey with Anatolian bees from North Africa that would erase the desire of his breed for propolis, P. Lavie in France showed that propolis had antimicrobial properties against the causative agent for American foulbrood, and thereby might be of benefit to bees. Lavie's work set off several decades of work, continuing now, aimed at characterizing specific propolis sources and specific bee diseases these sources might impact. More generally, propolis is one of many plant-derived substances that can impact bee health directly as antibiotics and indirectly by improving bee immune responses. Silvio Erler and Robin Moritz in Halle, Germany, have nicely summarized decades of experiments focused on determining the impacts of propolis and a wide pharmacopeia of bee-collected products on honey bee health (Erler & Moritz, 2016).

While health benefits to bees from propolis are evident from controlled experiments in the laboratory, studies of long-term effects on colony health are scarce. In one such study, Renata Borba and colleagues at the University of Minnesota placed commercial propolis traps into twelve experimental colonies, greatly increasing propolis levels in these colonies when compared to control colonies housed in standard hive bodies (Borba et al., 2015). These researchers then measured both disease loads and colony health across a full year. This experiment was repeated in consecutive years, involving a total of 24 experimental and 24 control colonies. In the first series, propolis-rich colonies survived better across winter and carried more brood when sampled the following spring. These colonies also invested less in six proteins known to be involved in the bee immune response during their first summer and fall, arguably conserving energy and resources. In the second round, propolis-rich bees showed reduced expression for only two of these six immune proteins, and no differences were found in either production or colony survivorship. Interestingly, bees in propolis-rich colonies showed much less variation for immune gene activity when compared to control bees, a result that might indicate a more uniform disease threat in these colonies, a good sign for bees threatened by a major colony-level disease breakout. Perhaps in support of this, *Nosema* loads were lower in propolis-rich colonies, albeit with marginal statistical confidence.

More recently, Nora Drescher and colleagues in Germany and Switzerland conducted an even more aggressive experimental manipulation of the propolis envelope (Drescher et al., 2017). As in the prior study, they installed commercial propolis traps in colonies. In half of these colonies, traps were removed frequently, scraping propolis and delivering it to one of the propolis-rich colonies. They also scraped the naturally stored propolis from the propolis-poor colonies, increasing the differences between each set of colonies. They then assessed disease levels and colony health monthly across a single growing season in a total of ten colonies. By the end of these assessments, propolis-rich colonies showed improved colony strength.

Neither study showed a significant effect of propolis on colony mite levels or on total virus or bacterial loads, in contrast to controlled lab studies. Interestingly, Drescher and colleagues showed that Deformed wing virus levels did not increase in step with their mite vectors in propolis-rich colonies, a hint that bees were in fact better able to fight off this virus. Finally, Borba and colleagues found differences in antibacterial activity when *Paenibacillus larvae* was exposed to propolis collected from colonies in the spring versus fall. Assuming both sets of propolis were derived from the same plant source, this result suggests that the benefits of propolis decrease in colonies over time.

These results indicate that the benefits of propolis can indeed outweigh the nuisance of sticky hive components. Natural propolis collected from multiple sources on two different continents was shown to lead to improved colony health and survival. It is exciting to think of the diversity of plant resins to which bees are exposed when collecting and unloading propolis. Surely, the observed antimicrobial traits seen in lab studies for distinct propolis sources can lead to natural or synthetic medicines for bees. Time will tell if beekeepers and scientists are better than bees in selecting specific plant medicines that consistently improve bee health.

Erler, S. & Moritz, R.F.A. (2016). Pharmacophagy and pharmacophory: mechanisms of self-medication and disease prevention in the honeybee colony (Apis mellifera). Apidologie, 47, 389-411. https://link.springer.com/article/10.1007/s13592-015-0400-z.

Borba, R.S., Klyczek, K.K., Mogen, K.L. & Spivak, M. (2015). Seasonal benefits of a natural propolis envelope to honey bee immunity and colony health. Journal of Experimental Biology, 218, 3689-3699. http://jeb.biologists.org/content/jexbio/218/22/3689.full.pdf.

Drescher, N., Klein, A.M., Neumann, P., Yañez, O. & Leonhardt, S.D. (2017). Inside honeybee hives: Impact of natural propolis on the ectoparasitic mite Varroa destructor and viruses. Insects, 8. doi:10.3390/insects8010015.

16

NEXT-GENERATION SCIENTISTS

I am a bit of a cynic for science fairs and award shows (and don't get me started on beauty contests), but I was deeply moved by the energy and insights provided by high school students from 81 countries at the 2018 Intel International Science and Engineering Fair (https://student.societyforscience.org/intel-isef). When my daughter Simone was invited to compete, I wasn't sure it would merit spending a week in Pittsburgh (and yes, I did joke that the second-place Maryland girl received TWO weeks in Pittsburgh), but fortunately Simone pushed, and off we went to the 'Olympics' of youth science.

The Fair included about 20 presentations related to pollination, and I tried to visit all of them. One favorite was presented by Elizabeth Wamsley from Timber Ridge Scholars Academy in Missouri. Elizabeth used RNA interference to target *Varroa* mites, a hot topic both in industry and at research labs, including ours. She was able to see an effect of RNA knockdowns after soaking mites in solutions containing RNA segments targeting *Varroa* mitochondrial proteins, broadening the list of potential mite targets. Elizabeth won scholarship offers and cash awards for her efforts.

Natalia Jacobson from Empire High School in Arizona also focused on a major honey bee health threat, in her case *Nosema* disease. Natalia provided new insights into the impacts of protein nutrition on *Nosema*. Specifically, by supplementing diseased bees with the amino acid cysteine, she measured increased immunity and survival along with the enlarged hypopharyngeal glands typical of healthy bees. As with other emerging research, "your results may vary," so please hold off on introducing cysteine into your colonies until more work is done. Still, Natalia's project and analysis were careful and the results are truly exciting.

ISEF researcher William Deering, JDE

On the medical side, William Deering from IDEA Homeschool in Alaska tested whether plant compounds added to a 'synthetic honey' comprised of sucrose syrup could lead to new antibiotics. After informing him of the unfortunate connotations of the term 'synthetic honey,' I leaned in to see which compounds he favored and to compare them to similar ongoing efforts aimed at bee health. William found that infusions based on extracts from Alaskan flowers led to lower bacterial growth in 16 out of 17 cases, indicating a wide potential for plant extracts as a new source of antibiotics. He is widening his scope to plant extracts from other parts of North America and beyond, and has the energy to really ramp up this search.

Also, in the 'what can bees do for *you*?' section, Australian student Ella Cuthbert from Lyneham High School had a brilliant project looking at medical uses for honey bee silk. To avoid some extremely tedious collections, she inserted the major silk proteins into a laboratory production system, allowing her to make these proteins in a test tube. In the end, two silk proteins were shown to reduce the growth of bacteria.

ISEF Researcher Ella Cuthbert, JDE

Combined with the strength characteristics of silk, this finding might lead to new wound dressings. Nobody defines youthful optimism better than Ella, who started her abstract with "The dream of living forever, only accomplishable by replacing broken parts of the human body, is closer than it has ever been before." Take THAT, curmudgeons.

Additional international entries came from Thailand (A New Method to Increase Propolis Production by Activating Nest Repair Behavior in Stingless Bees), South Africa (Determining the Availability of Pollen Sources for Honeybees on Deciduous Fruit Farms in Summer), and Puerto Rico (Comparative Study Between *Bellis sylvestris*, *Artemisia dracunculus* and *Lantana trifolia* in the Ability to Attract *Apis mellifera*). These entries, also, were well done and well presented.

ISEF team from Thailand, JDE

In the end, if I had to pick a 'Best in Show' it would go to Brooklyn Pardall from Central Lee High School in Iowa. Brooklyn showed a stunning increase in soybean yields when honey bees were part of the picture. Honey bees are not needed for soybean production, since production soy plants are self-pollinating, nor are bees really thought to crave the rewards

ISEF researcher Brooklyn Pardall, JDE

provided by soybean flowers. Nevertheless, having hive boxes amidst rows of soybeans seems to have increased yields by over 20%. This is a huge margin in a crop already pushed to production capacity. What gives? It could be that bee visits tap into a dormant plant reproductive cue, even when the pollen they deliver is no longer critical. In a similar vein, asexual coffee plants were shown by Smithsonian scientist David Roubik to respond favorably to honey bee visits, increasing bean yields substantially (Roubik, 2002). Brooklyn is planning follow-up experiments this year. Having met her and her team, and knowing that she carried out all of her experiments on her own family's 6,000-acre soybean farm, I would not bet against her. If it turns out that honey bees significantly improve soybean crops, this could be a three-way win; better crop production, a new revenue source for beekeepers, and a stronger incentive for soybean farmers to maintain a healthy environment for visiting pollinators.

Since I don't have a gig at a plant magazine, I can't give much space to Simone Evans' awesome orchid project nor her success at these Olympics, but she already knows who's the favorite…and despite my snarky comment at the start, Pittsburgh was really fun. It reminded me of the 1970's and 80's Seattle of my youth, before that city became so sparkly. If you are able, I would strongly recommend attending the public days at INTEL/ISEF in the future to see some great ideas and outputs (Phoenix in 2019, Anaheim in 2020). You can also peruse projects from 2018 and many former years at https://abstracts.societyforscience.org/. Many past competitors continue to have a great impact on the field. Regardless, please support your local science students, relatives or not, as they set out on ways to improve the world.

Society for the Science & the Public. International Science and Engineering Fair (ISEF). Then New Generation of Innovators. Received at https://student.societyforscience.org/intel-isef.

Roubik, D.W. (2002). The value of bees to the coffee harvest. Nature, 417. https://www.nature.com/articles/417708a.

Society for the Science & the Public. Abstract Search. Received at https://abstracts.societyforscience.org/.

17

LOVE IN THE TIME OF CHASMOGAMY: HOW AND WHEN DO BEES IMPROVE SOYBEAN YIELDS?

As I wrote in the previous essay, the Intel International Science and Engineering Fair (https://student.societyforscience.org/intel-isef) had plenty of optimistic news for bees and their roles in nature and agriculture. Brooklyn Pardall (Central Lee High School, Iowa) highlighted research she carried out on her family's 6,000-acre soybean farm, showing significant benefits from the exposure of production soybean plants to honey bees. Soybeans represent a crop where the benefits of insect pollinators are complex and surely underestimated. In theory, soybeans could provide a ubiquitous resource for beekeepers seeking new flowers and profits, assuming growers in fact see an increase in yields when bees (wild or managed) are on the scene. Here, I delve into the long history of research on insect pollination and soybean yields. Like a lot of research, this topic sometimes feels like déjà vu all over again, but it is clear that more could be done to advertise soybean cultivars and conditions that benefit from bee pollination and to incentivize beekeepers to take the plunge into a widespread floral source. Given hard evidence, soybean growers might choose cultivars and safe management techniques that reward bees and beekeepers. Soybeans cover 90 million acres in the U.S. (http://usda.mannlib.cornell.edu/usda/current/CropProdSu/CropProdSu-01-12-2018.pdf), so if even a fraction of this landscape joins the pollen-nation that seems like a potentially big deal.

A first stop on any pollination quest, especially for a non-specialist like me, is USDA researcher S.E. McGregor's 'Insect Pollination of Cultivated Crop Plants' (1976, https://www.ars.usda.gov/ARSUserFiles/20220500/OnlinePollination Handbook.pdf). McGregor was not big on bee pollination of soybeans to say the least, stating "the soybean is considered to be self-fertile and not benefited by insect pollination," and "there are no recommendations for the use of bees in pollination of soybeans." Still, he does leave the door open a bit with, "Although there is no experimental evidence to support them, some soybean growers in Arkansas have indicated that bees increase production of beans, and they encourage the presence of apiaries near their fields." And my favorite sign of smoke, amidst a discussion of how outcrossing 'could' improve yields, is "various tests have been

Soybean field, YH

conducted to determine the amount of cross-pollination that occurred at different locations, but the agents responsible for the crossing obtained were usually not determined, possibly because the tests were conducted by agronomists who did not consider themselves qualified to record these observations."

Dr. Eric Erickson apparently ignored his USDA colleague's pessimism and set off at the same time on a research program to measure the impacts of honey bees on soybean yields. With collaborators including a young Kim Flottum, Dr. Erickson mapped out some of the benefits of bee pollination to soybean yields in the mid-1970's, beginning with "Effects of honey bees on yield of three soybean cultivars" (Erickson, 1975). Dr. Erickson provided experimental data from the field to counter the dogma that bees have a minimal impact on soybean yields. As hinted by the title, not all soybean varieties attract bees nor do they all provide good forage for bees. The 'Hark' variety was the most chasmogamous ('open marriage' for insect pollinators) and indeed plants of this variety showed 7% and

16% yield increases when exposed to bees, in successive years. Two varieties with closed flowers (cleistogamous) showed no benefit from bees and few visitors. Wainer Chiari and colleagues also showed that varieties make the difference, with bees (nearly always honey bees in their study) boosting productivity of Brazilian cultivar BRS-133 by over 50% (Chiari et al., 2005). Presumably, BRS-133 is a variety that thrives on cross-pollination and its associated influx of genetic diversity. Finally, Diego Blettler and colleagues in Argentina recently showed an 18% yield increase when soybeans were exposed to insects in one year, but no impact for the same soybean cultivar the second year (Blettler et al., 2018). The honey bee chauvinist in me says these benefits point to the immense roles of honey bees in crop production, but there is good evidence that other bees are playing significant roles in increasing the soybean crop. Marcelo de O. Milfont and colleagues in Brazil found that native bees alone boosted soybean yields by 6% while honey bees added 18% on top of that (Higher soybean production using honeybee and wild pollinators, a sustainable alternative to pesticides and auto-pollination, (de O. Milfont et al., 2013)). Recent work in the U.S. suggest that multiple pollinators can still be found in soybean fields (e.g., Defining the insect pollinator community found in Iowa corn and soybean fields: Implications for pollinator conservation, by Matthew O'Neal and colleagues (Wheelock et al., 2016)), and these pollinators are likely to enable gene flow in soybeans alongside honey bees.

Despite the observed benefits from these and other studies, growers will not be convinced to welcome bees until there is more clarity on specific cultivars that benefit from pollination and the conditions needed for strong bee-enabled 'marriages'. Surprisingly given its role as the leading row crop in many countries, there does not seem to be a concerted effort to identify specific soybean varieties that benefit from the gene flow provided by insect pollinators, nor are there steady recommendations for pollinator density in production soybeans.

So, what is the big picture for U.S. soybean production, and can the soybean crop be made to be more bee-dependent, and bee-friendly? Nicholas Calderone produced an ambitious summary of the impacts of honey bees and other pollinators on U.S. crop production ("Insect Pollinated Crops, Insect Pollinators and US Agriculture: Trend Analysis of Aggregate Data for the Period 1992–2009," (Calderone, 2012)), as a follow-up to work by himself and Roger Morse in 2000. Using USDA data, he calculates acreages and productivity for a range of commodities that depend on, or benefit to some extent from, pollinators. Soybeans make this list, albeit with an estimated 10% boost from insect pollination. This 10% is no-doubt an informed choice but it misses the nuances of which varieties are 'open' to bees, which planting schemes favor pollination and gene flow, and which soybean management practices are at least somewhat bee friendly. Ten percent of 90 million acres is still substantial, and for 2010, Calderone estimates that bees provide nearly $4 billion in value for soybean growers, arbitrarily split between honey bees and all other pollinators. This surpasses the $2.8B in pollinator benefits for almonds in that year, effectively all from honey bees despite inroads recently from alternate managed or wild pollinators. To be sure, almonds are a high-value crop grown on 1% of the acreage used by soy, but if conditions could be worked out between soybean growers and beekeepers that provide growers substantial benefits and beekeepers some profit and safe forage there seems to be plenty of space for some chasmogamy. If so, the incentive and insights will come from researchers such as Ms. Pardall who know and care for both sides of this plant-pollinator relationship.

Society for the Science & the Public. International Science and Engineering Fair (ISEF). Then New Generation of Innovators. Received at https://student.societyforscience.org/intel-isef.

http://usda.mannlib.cornell.edu/usda/current/CropProdSu/CropProdSu-01-12-2018.pdf.

United States Department of Agriculture. Insect Pollination of Cultivated Crop Plants. The First and Only Virtual Beekeeping Book Updated Continuously. Received at https://www.ars.usda.gov/ARSUserFiles/20220500/OnlinePollinationHandbook.pdf.

Erickson, E.H. (1975). Effect of Honey Bees on Yield of Three Soybean Cultivars. Crop Science, 15, 84-86. https://dl.sciencesocieties.org/publications/cs/abstracts/15/1/CS0150010084.

Chiari, W.C., De Toledo, V.D.A.A., Ruvolo-Takasusuki, M.C.C., Braz De Oliveira, A.J., Sakaguti, E.S., Attencia, V.M. et al. (2005). Pollination of Soybean (Glycine max L. Merril) by Honeybees (Apis mellifera L.). Brazilian Archives of Biology and Technology, 48, 31-36, http://www.scielo.br/pdf/babt/v48n1/a05v48n1.pdf.

Blettler, D.C., Fagúndez, G.A. & Caviglia, O.P. (2018). Contribution of honeybees to soybean yield. Apidologie, 49, 101-111. https://link.springer.com/article/10.1007/s13592-017-0532-4.

de O. Milfont, M., Rocha, E.E.M., Lima, A.O.N. & Freitas, B.M. (2013). Higher soybean production using honeybee and wild pollinators, a sustainable alternative to pesticides and auto pollination. Environmental Chemistry Letters, 11, 335-341. https://link.springer.com/article/10.1007/s10311-013-0412-8.

Wheelock, M.J., Rey, K.P. & O'Neal, M.E. (2016). Defining the Insect Pollinator Community Found in Iowa Corn and Soybean Fields: Implications for Pollinator Conservation. Environmental Entomology, 45, 1099-1106. https://academic.oup.com/ee/article/45/5/1099/2197225.

Calderone, N.W. (2012). Insect Pollinated Crops, Insect Pollinators and US Agriculture: Trend Analysis of Aggregate Data for the Period 1992–2009. PLoS ONE, 7. https://doi.org/10.1371/journal.pone.0037235.

PART THREE

ROYAL DECREES

18

REPLACEMENT QUEENS, A TRUE CINDERELLA STORY

Honey bee colonies and beekeepers alike can't afford to take queen rearing lightly. During supercedure events, swarm preparation, or emergency queen replacement, nurse bees must reach a quick consensus on which female larvae will receive royal care. Two recent articles from leading bee scientists offer contrasting views for how that choice is made. Ramesh Sagili and colleagues suggest that nurse bees make an economic choice, by tending to invest in well-fed young larvae as their future queens. Presumably, young female larvae that are large for their age will develop into more robust queens. Sagili and colleagues starved clusters of first-instar female larvae for four hours by blocking out nurse bees with a mesh screen. They then placed combs containing these clusters, alongside clusters of larvae that had been fed normally, into colonies that were ready to raise a replacement queen. Food-deprived larvae, while still viable, were far less likely to be picked as replacement queens. Interestingly, the discrimination against skinnier larvae was not observed when starved and normal larvae were placed into queen cups, suggesting that the stimulus of these cups outweighed any cues nurses might use to pick the most queen-worthy larvae. As further evidence that nurse bees can smell starvation, or that starving larvae tend to beg more for food (or both), larvae that had been deprived of food were fed more often and for longer periods. Thus, nurse bees did their best to get these larvae caught up in terms of worker development, despite tending to pass them over as future queens. Their work is described in "Honey bees consider larval nutritional status rather than genetic relatedness when selecting larvae for emergency queen rearing" (Sagili et al., 2018).

This and other experiments suggest that physical cues and/or chance are the main forces driving which larvae are chosen for the queen route, but this need not be the case. Much thought and work have gone into testing whether there might be a birthright for royalty. This could come in two ways. First, since queen bees mate prolifically, most females in the nest are only half-sisters. If a nurse bee had the power to choose a full sister as the next queen, this nepotistic act would benefit her greatly (especially since the dads of full or 'super'-sisters are haploid and hence give an exact version of themselves to all of their daughters, a tale for another day). As appealing as nepotism seems, it is not a frequent occurrence in bee colonies, arguably because a nepotistic trait would lead to poor queen

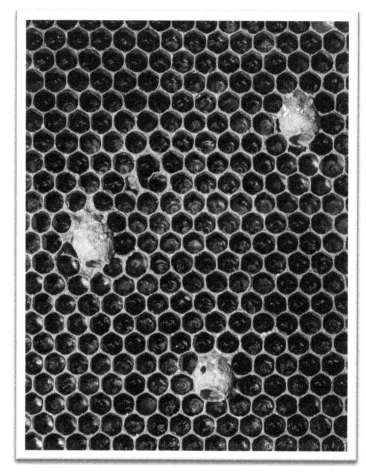

Queen cells, HFB

regulation over time, if not outright battle. There is another way for a genetic bias to present itself. Perhaps there are blue-blooded lineages of honey bees whose members inevitably aim for a royal seat. This, too, can lead to an unstable state where successful royalty begets more royalty until the ratio of queen-destined larvae gets out of whack, at its extreme, leading to dozens of feuding royals and not a lot of stinging or foraging. Still, when this trait is rare there is an opportunity for such royalty genes to hang around.

James Withrow and David Tarpy present strong evidence for 'royalty' lineages in honey bees. Using a form of DNA fingerprinting, they show a tendency for some lineages to be over-represented as queens in their article, "Cryptic 'royal' subfamilies in honey bee (*Apis mellifera*) colonies" (Withrow & Tarpy, 2018). Chosen royals are extremely rare. In fact, the authors propose that queen bees are even more promiscuous than currently thought, and some of the most rare lineages (patrilines) show this royal bias. So how does this trait arise and why don't royal lineages predominate? Mechanistically, royal-leaning larvae must attract special attention from nurse bees, perhaps like the starved larvae above, but without going through starvation. Since nepotism is not involved, these 'feed-me-well' cues must be perceived by average nurse bees. The authors argue that the other forms of queen replacement, namely supercedure and pre-swarming, slow the rise of royal lines. Here, queen-destined larvae are laid in preformed queen cells and it seems implausible that a queen aspirant could determine her own birth order in such a way as to land in a queen cell.

A true fairy tale skips over the gritty details, but if you are interested in which changes occur *inside* chosen queens when they start on a royal path, numerous scientists have tackled this. One way involves measuring how larvae turn on specific genes that lead to the proteins best suited for either royalty or a life of work. Xu-Jiang He and colleagues have provided a complete view of these caste-biased genes in their recent paper, "A comparison of honeybee (*Apis mellifera*) queen, worker and drone larvae by RNA-Seq" (He et al., 2017). This topic is fascinating to me personally and, in fact, is the very question that brought me into bees 20+ years ago. Diana Wheeler, Gloria deGrandi-Hoffmann and I recount the ways specific genes can lead to queens or workers in "Honey bee queen production: Tight genes or too much food?" (Evans et al., 2000). The next time you induce a queen event in your colonies, or watch as one unfolds, remember the many forces inside new queens and their nestmates that can affect the outcome.

Sagili, R.R., Metz, B.N., Lucas, H.M., Chakrabarti, P. & Breece, C.R. (2018). Honey bees consider larval nutritional status rather than genetic relatedness when selecting larvae for emergency queen rearing. Scientific Reports, 8. https://www.nature.com/articles/s41598-018-25976-7.

Withrow, J.M. & Tarpy, D.R. (2018). Cryptic "royal" subfamilies in honey bee (Apis mellifera) colonies. PLoS ONE, 13. https://doi.org/10.1371/journal.pone.0199124.

He, X.J., Jiang, W.J., Zhou, M., Barron, A.B. & Zeng, Z.J. (2017). A comparison of honeybee (Apis mellifera) queen, worker and drone larvae by RNA-Seq. Insect Science. https://onlinelibrary.wiley.com/doi/full/10.1111/1744-7917.12557.

Evans, J., Degrandi-Hoffman, G. & Wheeler, D. (2000). Honey bee queen production: Tight genes or too much food? American Bee Journal, 140, 136-137.

19

HISTORY TENDS TO REPEAT ITSELF

It can be easy to forget when working with honey bees that our human existence as their keeper's pales with respect to the time they kept themselves going just fine without us. Like other highly social insects, honey bees were widespread and important players in nature for millions of years before humans recognized them as a partner. This is not to understate the human:bee relationship; we have been friends with benefits for thousands of years. Instead, picture the many millions of years honey bees have spent landing on flowers of plant species still found today. During all that time, honey bees faced diverse climate zones and seasonality similar to today's. They battled at least some of today's parasites and pathogens, and harbored versions of today's symbiotic microbes. These relationships with the living and physical world pushed bees into most of the behaviors and physical traits we see now. In other words, this evolutionary history set the stage for many of the strengths and weaknesses of honey bees as a species, baggage we humans both exploit and struggle to 'manage.'

Previously, I highlighted Tom Seeley's work on the biologies of successful feral and 'alt-feral' honey bee colonies. While I have long enjoyed Tom's work and could write about it monthly, I was inspired to do that column largely thanks to a fascinating workshop Tom helped lead (with Professors Mark Winston, Marla Spivak, and an irrepressible beekeeping couple from California, Bonnie and Gary Morse, among others). This "Bee Audacious" workshop dug into some truly unique ideas for changing the beekeeping world, from Darwinian beekeeping to bees as spiritual guides (truly all sides were at the table and in respectful conversation). A key point of these discussions was that we ignore the long history of honey bees, their food sources, and their environment at our peril.

Evolutionary thinking with respect to bees is not completely novel. Charles Darwin mused about both flowers and social insects as he built the case for evolution. More recently, Marla Spivak brought many of us to our feet with her plea for '(r)evolutionary beekeeping' almost two decades ago. What WAS novel and audacious was the effort to bring beekeepers, producers, and scientists into a secluded place and turn the screws to make them come up with some practical outcomes. A living conversation of this effort is on the web at Beeaudacious.com, including an extensive review drafted by Mark Winston and Nicole Armos from Simon Fraser University (http://beeaudacious.com/wp-content/uploads/2017/04/BA_Final_Reportv1.8.2opt.pdf).

In a similar vein, bee researchers Berry Brosi and Keith Delaplane along with two disease-expert colleagues have just published an excellent overview of evolutionary thinking as applied to honey bee disease. This review is packed with insights into why mites and viruses are bad and getting worse for bees. In particular, management practices from colony crowding to shared equipment and imperfect disease controls can actually drive parasites such as mites and viruses to fight back with increased virulence as they race to beat our controls. As one example, periodic yet ineffective treatments for mites can favor those mites that grow fastest in between applications. Racing the clock by parasites often leads to higher virulence or impacts on their hosts. This is on top of the frustrating evolution of resistance to the treatments themselves, a race that has us constantly searching for the next big thing in mite control. The review is behind a pay wall at *Nature Ecology and Evolution* (Brosi et al., 2017), but if

Bee frame health inspection, HFB

you contact the authors they can likely give you substantial details from this important paper.

How can beekeepers and those who want bees and their fruits take advantage of history and the dynamic field of evolutionary biology? One way is to be open to change. Listen to both the young and old beekeepers in your circles that are trying new ways to manage colonies. Maybe this involves new approaches to swarming and splits, overwintering, queen replacement, and pest control. Maybe it involves tolerating an adverse trait that has an ancient record of helping bees (like sticky propolis?). For researchers, this might mean being open to test some of the more exotic new treatments for disease or nutrition. These are often derisively called 'fairy dust,' although I prefer 'magic potions' since there is a human hand in their genesis, even if it feels like alchemy sometimes. One or more of these products will indeed provide a safe and effective aid to bees and many people are hard at work trying to identify those scarce winners. They are not a quiet bunch in general so you have likely seen this. The groups with the best chance to identify 'out of the blue' winners have a firm grasp on the evolution of their plant, microbial, and even mushroom sources and they are using these insights to knock down nature's pharmacopeia to a size that works for real testing.

Evolutionary thinking can also help explain the vexing sensitivity of bees to agro-chemicals and to the rare plant nectars containing toxic compounds. As a needed partner to flowering plants, bees have historically received 'the good stuff,' and their physiologies seem to reflect a naiveté toward chemical insults. One of the discoveries of the honey bee genome project was that bees indeed have fewer proteins known for detoxifying chemicals than do other insects (Claudianos et al., 2006). Finally, to improve honey bees in the long term, breeders must build on existing traits controlled by a mess of intersecting chemical roads and protein pathways. To bring out the best traits takes care, since other less desired genetic traits can come along for the ride or can slow down the spread of these winners. Since many genes need to be juggled to keep winning traits around, it is also incredibly hard to hang on to what you have in a breeding program. The best breeders will focus on traits that both respond to genetic pushes (e.g., are heritable) and are still expressed when the entropy of the real world hits them. In breeding and bee management, we ignore history at our peril.

Beeaudacious. Audacious Visions for the Future of Bees, Beekeeping and Pollination. Received at http://beeaudacious.com/wp-content/uploads/2017/04/BA_Final_Reportv1.8.2opt.pdf.

Brosi, B.J., Delaplane, K.S., Boots, M. & De Roode, J.C. (2017). Ecological and evolutionary approaches to managing honeybee disease. Nature Ecology and Evolution, 1, 1250-1262. https://www.nature.com/articles/s41559-017-0246-z.

Claudianos, C., Ranson, H., Johnson, R.M., Biswas, S., Schuler, M.A., Berenbaum, M.R. et al. (2006). A deficit of detoxification enzymes: Pesticide sensitivity and environmental response in the honeybee. Insect Molecular Biology, 15, 615-63. 10.1111/j.1365-2583.2006.00672.x.

20

CONVERGENT WAYS TO EXPOSE AND FIGHT MITES

I have highlighted the challenges needed to select for and maintain specific desired traits in honey bees while not losing ground on others, e.g., work on Varroa-sensitive hygiene (VSH) and other resistance traits (Essay 19). Recent months have seen at least four advances in identifying and selecting these desirable traits. One study compiled insights from four of the best known Varroa-resistant populations, looking for common qualities that arise and are maintained when bees are either actively bred for low mite loads or develop mite resistance naturally. These include a sustainable mite population in Norway, Sweden (Gotland) and France. In all cases, a behavioral component appears to help these lineages keep mite levels in check. Melissa Oddie and colleagues in their article, "Rapid parallel evolution overcomes global honey bee parasite," take one aspect of hygienic behavior and isolates it from similar defenses (Oddie et al., 2018). The trait they chose was the frequency of cell uncapping and recapping, without removing brood. Interestingly, simply uncapping the cells and letting bees recap them led to a substantial increase in non-reproductive mites. While they

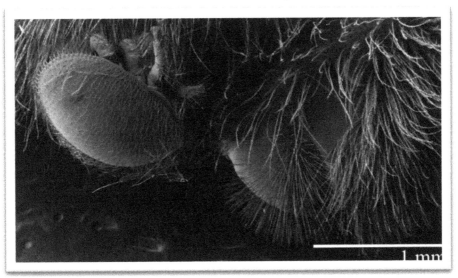

Varroa mite on worker bee, GB

used a creative method to do this in quantity (soaking a piece of linen with wax, letting it adhere to caps and then ripping the lids off), this is not a method beekeepers will do much. Still, it gives insights into the benefits of uncapping, per se. As another creative tool, they recognized that cells that had been uncapped at some point during development showed a characteristic dimple that could be seen and documented at the very end of development. This neat trait allowed them to simply screen mature pupae for mite loads and then infer whether they had been uncapped at some point (saving the eyes of countless students who would otherwise have to stare at an observation hive with capped brood). As expected, the four resistant stocks showed significantly higher recapping rates during development, 3-4-fold more often than the susceptible stocks. Recapping was focused on mite-infested brood, but even non-infested brood was recapped at a higher rate in resistant stock, suggesting a general tendency to look under the hood, or perhaps the presence of other diseases that triggered hygienic behavior. Importantly, the uncapped brood used in these experiments came from a homogeneous set of donor colonies. In other words, it was not that the diseased brood in hygienic colonies were yelling (smelling) louder, but that the workers uncapping and then recapping their cells were somehow more attuned and active. Perhaps the recapping 'dimple' can be used by bee breeders as another strategy to breed resistant stock.

Hasan Al Toufailia and colleagues in England recently confirmed experimentally that a key for honey bee hygienics is the ability to recognize what is going on with capped brood in their article, "Both hygienic and non-hygienic honeybee, *Apis mellifera*, colonies remove dead and diseased larvae from open brood cells" (Toufailia et al., 2018). In their study, worker bees removed all open brood that was freeze-killed within a day, in all lineages of bees studied. This was true across a set of 20 colonies showing a wide spectrum of hygienic tendencies (53%-100% removal of freeze-killed and sealed brood). Similarly, when the youngest larvae were exposed to *Ascosphaera apis*, the causative agent for chalkbrood disease, all exposed larvae were removed pre-capping. In contrast, when larvae were exposed to *A. apis* closer to capping, only around 30% ('medium-aged' larvae) and 15 (larger larvae inoculated a day before capping) were uncapped and removed by workers. Some of these differences in brood removal might reflect resilience or ignore capped disease larvae. So, the hunt needs to focus on what makes nurse bees more attuned to their stressed younger sisters while those sisters are covered by a layer wax.

Seo Hyun Kim and colleagues made headway on determining the cues hygienic worker bees recognize in diseased-capped brood. In their 2018 study, "Honey bees performing *Varroa* sensitive hygiene remove the most mite-compromised bees from highly infested patches of brood" (Kim et al., 2018), bees were more likely to uncap brood to check things out when mite foundresses were not only reproductive but also actively so. This has been demonstrated as a key trait of VSH in the past (see work by Marla Spivak and colleagues in the *Journal of Neurobiology*, 2003, (Spivak et al., 2003)), and the current study pushes the science forward by showing exactly when, and perhaps how, mite reproduction triggers uncapping. Cells with mite offspring, even tiny protonymphs, were uncapped twice as often as cells with just a mite foundress or a foundress with eggs. Once a cell was uncapped, it was more likely that neighboring cells would be uncapped even when accounting for their own mite levels, so bees seemed to be accurately predicting that mites cluster somewhat in small regions of the comb.

Finally, Alison McAfee and colleagues in Vancouver quantified the abilities of two volatile chemicals, β-ocimene and oleic acid, to trigger hygienic behavior by worker bees (McAfee et al., 2018). While it is not perceived by our noses, oleic acid is a widely used indicator of death throughout the arthropods (insects, crustaceans, mites and the like), meaning that members of these groups have been using this cue to avoid their own demise for 400+ million years (Yao et al., 2009). Other than confirming once again that death stinks, how can these results be used to advance bee breeding? One way is to recognize exactly how bees perceive these two molecules and then determine whether this mechanism can be enhanced via breeding. McAfee and colleagues are well on the way to doing just that by exploiting a remarkable set of specific proteins that help bees and all of us smell our environment. These aptly named odorant binding proteins (OBPs) are diverse in honey bees and are involved in many aspects of their communication, often being triggered by single molecules that fit them just right. Two OBPs, OBP16 and OBP18, seem to react to β-ocimene and oleic acid and hence are targets for breeding more perceptive bees. The discovery that decades of excellent work on hygienic behavior in honey bees can be refined to specific cues is truly exciting and these insights should aid breeding efforts against some of the worst honey bee foes.

Oddie, M., Büchler, R., Dahle, B., Kovacic, M., Le Conte, Y., Locke, B. et al. (2018). Rapid parallel evolution overcomes global honey bee parasite. Scientific Reports, 8.

Toufailia, H.A., Evison, S.E.F., Hughes, W.O.H. & Ratnieks, F.L.W. (2018). Both hygienic and non-hygienic honeybee, Apis mellifera, colonies remove dead and diseased larvae from open brood cells. Philosophical Transactions of the Royal Society B: Biological Sciences, 373. DOI: 10.1098/rstb.2017.0201.

Kim, S.H., Mondet, F., Hervé, M. & Mercer, A. (2018). Honey bees performing Varroa sensitive hygiene remove the most mite-compromised bees from highly infested patches of brood. Apidologie, 49, 335-345. DOI: 10.1007/s13592-017-0559-6.

Spivak, M., Masterman, R., Roco, R. & Mesce, K.A. (2003). Hygienic behavior in the honey bee (Apis mellifera L.) and the modulatory role of octopamine. Journal of Neurobiology, 55. DOI: 10.1002/neu.10219.

McAfee, A., et al. (2018). "A death pheromone, oleic acid, triggers hygienic behavior in honey bees (Apis mellifera L.)." Scientific Reports 8(1).

Yao, M., Rosenfeld, J., Attridge, S., Sidhu, S., Aksenov, V. & Rollo, C.D. (2009). The Ancient Chemistry of Avoiding Risks of Predation and Disease. Evolutionary Biology, 36, 267-281. DOI: 10.1007/s11692-009-9069-4.

21

HOLDING THE LINE ON TRAIT ROT AND INBREEDING

As a geneticist, I tirelessly support (mostly from a safe distance) programs to improve the resilience of bees through breeding. History has shown that honey bees contain a continuum of behaviors, defenses, and colony-level traits. This variation has been exploited by beekeepers for decades, in hopes of producing queens, drones, workers, and colonies that are best suited to local climates and show desired survival, behavior or productivity traits. Much of the focus of serious breeding programs is on resistance to *Varroa* mites or tolerance of these mites when they are in the hive. There is good news on that front, from survivor stock that limits mite fertility (e.g., work on Norwegian bees by Melissa Oddie and colleagues (Oddie et al., 2017) to stock with *Varroa*-sensitive hygiene (VSH), e.g., Robert Danka, Jeff Harris, and their colleagues at USDA-ARS and Mississippi State University, respectively).

It is hard enough to identify and breed from desired traits, but honey bee breeding efforts also suffer from difficulties in hanging on to those same traits. This phenomenon, which I would call "trait rot," is relentless. Even the best traits get reshuffled each generation, meeting up with additional genes and combinations of genes. In general, these hookups cause a trait that was initially extreme (like ultra-hygienic behavior) to become diluted in subsequent offspring. At the colony level, desired traits found in worker bees can be further diluted by the sheer diversity of genes across the colony (the "many-dad" syndrome). Unless selection is intense or mating is closely controlled these processes can, and generally do, dampen the strength of initial breeding traits. Were this not the case, breeder queens would not be so pricey relative to their many daughters.

Selection for disease resistance has an added cost in that those being selected against (parasites and pathogens) have their own motives for defeating each resistance trait sent their way. The so-called Red-Queen hypothesis (reflecting the Lewis Carroll character who spends her every minute "running, running, just to stay in the same place") defines this challenge to both human-driven and natural selection. In short, parasites and pathogens are in an arms race with their bee targets and can be quick to adapt to bee defenses. In the case of *Varroa*, selection for mites that beat new host defenses mirrors selection on these same mites to defeat the many acaricides used by beekeepers.

Finally, true genetic winners, whether this is accomplished in nature or with a human hand, can be successful so quickly that they abandon genetic variation that is equally important for the long-term survival of bees. This can lead to direct inbreeding costs (diploid males, for one) or subtle losses of genetic variants that could be needed in the future. One example of the latter would be an immune trait for a periodic disease (chalkbrood, viruses, etc.) that was not present in the population under selective breeding. Bees bred from a closed population, until their genes are vetted and merged into the wider gene pool, could have unforeseen but important weaknesses. This is but one downside of genetically modified bees; if they are overwhelmingly successful against, say, *Varroa* or the viruses tied to *Varroa*, they run the risk of sweeping through bee populations so quickly that millennia of genetic variation is lost.

Given this backsliding, breeders must identify and promote traits that will be robust and available to beekeepers for multiple generations. Writing in *Apidologie* (Danka et al., 2016), Danka, Harris, and Garrett Dodds show one approach for assessing backsliding, namely screening next-generation bees to see how closely their behaviors match the parent stock. In the case of VSH-selected stock, the authors claim this trait holds up well for one generation of outcrossing, in terms of both the removal of mite-infested brood and total mite loads. Lelania Bourgeois and Lorraine Beaman (also at the USDA-ARS Honey Bee Breeding, Genetics and Physiology Laboratory in Baton Rouge, Louisiana) used a genetic approach to address trait rot with respect to longstanding efforts to breed disease-resistant Russian honey bees described in their article, "Tracking the genetic stability of a honey bee (Hymenoptera: Apidae) breeding program with genetic markers" (Bourgeois & Beaman, 2017). They first identified a genetic signature for Russian breeding stock using 'chords' composed of eight genetic notes. This signature correctly distinguished individual Russian bees from the rest of the population about 80% of the time in the first year of release. While this percentage dipped the first year after release, it has since held steady, increasing confidence in stock integrity. These same genetic tests are also being used to assess inbreeding levels over time, providing a warning when populations become a bit too closed. Insights into the inbreeding costs of intense selection can be gleaned from natural experiments as well. In one study that marries multi-year fieldwork with genetics, Michael Lattorff and colleagues were able to show the famous 'Gotland' survivor stock in Sweden lost a substantial amount of genetic diversity between 2000 and 2007. Interestingly, specific regions of the 'Gotland' genome showed exceptionally low diversity, a clue these scientists are using to identify genes that were especially successful, effectively 'sweeping' neighbors on their chromosomes to high frequencies at the expense of other variants. Their study, "A selective sweep in a *Varroa destructor* resistant honeybee (*Apis mellifera*) population" is freely available (Lattorff et al., 2015).

Bee approaching flower, JE

Emerging genome-level resources for honey bees promise to speed efforts to both identify key *Varroa* resistance traits and keep them in the population. These resources include complete genome sequences for many bee species. Comparing across genomes should identify genetic signatures of bees that present (*Apis cerana*, e.g, a study by Zheguang Lin and colleagues, (Lin et al., 2018)) or lack natural roadblocks to *Varroa* reproduction.

Oddie, M.A.Y., Dahle, B. & Neumann, P. (2017). Norwegian honey bees surviving Varroa destructor mite infestations by means of natural selection. PeerJ, 2017. DOI 10.7717/peerj.3956.

Danka, R.G., Harris, J.W. & Dodds, G.E. (2016). Selection of VSH-derived "Pol-line" honey bees and evaluation of their Varroa-resistance characteristics. Apidologie, 47, 483-490. DOI: 10.1007/s13592-015-0413-7.

Bourgeois, L. & Beaman, L. (2017). Tracking the Genetic Stability of a Honey Bee (Hymenoptera: Apidae) Breeding Program with Genetic Markers. Journal of economic entomology, 110, 1419-1423. DOI 10.1093/jee/tox175.

Lattorff, H.M.G., Buchholz, J., Fries, I. & Moritz, R.F.A. (2015). A selective sweep in a Varroa destructor resistant honeybee (Apis mellifera) population. Infection, Genetics and Evolution, 31, 169-176. https://doi.org/10.1016/j.meegid.2015.01.025.

Lin, Z., Qin, Y., Page, P., Wang, S., Li, L., Wen, Z. et al. (2018). Reproduction of parasitic mites Varroa destructor in original and new honeybee hosts. Ecology and Evolution, 8, 2135-2145. DOI: 10.1002/ece3.3802.

22

MAGIC BULLETS FOR MITES

Varroa mites remain enemy number 1 among the living threats to honey bees. This is despite decades of attempts to understand and control these mites in Europe, then the Americas, and now across most of the beekeeping world. A prominent bee research professor told me in the early 2000's that the *Varroa* mite issue was under control, new chemical treatments were doing the trick and the mite would soon be vanquished. This was comforting to me at the time, since I had been hired to research American foulbrood disease. I was beginning to suspect that AFB was a 1950's sort of issue for beekeepers, who had just received word that small hive beetles were spreading and who seemed quite focused on mite impacts. Almost 20 years later, that professor has long retired and *Varroa* mites did not follow his suggestion.

A key strategy for reducing *Varroa* impacts is to interfere with female mating and reproduction. Coupled with hygienic behavior and controls aimed at vulnerable, exposed, female mites, reducing female reproductive success can greatly reduce the threats of *Varroa* mites at the colony level. This is due to the fact that *Varroa* females really aren't that productive, rearing at best a few female offspring in each reproductive round and generally, at least in worker cells, producing just a couple new mites. If reproduction can be reduced by even by a fraction, this will help cap mite

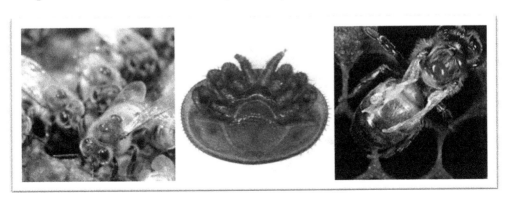

Varroa mites and their aftermath, ARS image gallery

levels late in the season, when mites become filthy circulators of bee viruses.

Benjamin Conlon and colleagues have made a major breakthrough in the identification of potential features of bee physiology that put a lid on mite reproduction. In their study, "A gene for resistance to the *Varroa* mite (Acari) in honey bee (*Apis mellifera*) pupae" (Conlon et al., 2019), the authors used a mapping strategy to identify a high-value target for bee breeding.

As with most scientific breakthroughs, the study required luck, insight, and collaboration across several research groups. The luck came in the form of a honey bee population in Toulouse, France, that showed an extremely high frequency of failed mite reproduction. The insight came in only screening male honey bees from that population. Besides just targeting the sex that is most attractive and supportive of *Varroa*, the focus on males allowed these researchers to use a fascinating trait of honey bee biology to speed the search for genetic resistance. Male honey bees are haploid from birth to death. Unlike female queens and workers (and ourselves for that matter) male honey bees have no father and every cell in their bodies holds chromosomes that are a genetically identical copy of the egg that gave rise to that individual. This is golden for genetic trait analysis, since traits cannot be masked by variants on an alternate chromosome and as such are fully exposed to view. It is also golden for breeding since, once identified in a single male, a trait can breed in an invariant way into every queen and worker who that male fathers (through the wonder of instrumental insemination). Double golden and a rarely exploited tool for bee breeding.

More luck came in finding a single colony in which half of the males were conducive to mite reproduction while half stymied female mites. For the latter half, all female mites emerged, if they emerged at all, with no offspring. If you remember Mendel and his peas, or some other lesson from a distant genetics class, this perfect 50/50 ratio suggests that the trait of interest, here a 'something' that keeps mites from starting a family, is most likely driven by a single gene. Triple golden, since forcing a single gene variant into a population and keeping it there by breeding is way easier than juggling dozens of genes each of which plays a minor role, the bugbear of breeding schemes in bees, plants, and livestock.

One last fact of bee biology worked to the advantage of these authors. Across most of the tree of life, when a female generates the haploid (single set of chromosomes) eggs that will be fertilized with a male-sourced similarly haploid cell (sperm, pollen, etc.) to make a diploid (two-copy) offspring, there is a substantial amount of sloppiness between the two pairs of chromosomes that give rise to these haploid sex starters. This leads to crossing over (recombination) and a general reshuffling of the paired decks of genes held by both mom and dad. Reshuffled decks help expose new combinations of traits to selection, a phenomenon thought to be the key driving force for sex in the first place (don't worry I will mention hormones too, below). If you find this primer on egg production confusing, or boring, just remember one thing, honey bee eggs reshuffle their decks at a rate tenfold higher than almost all other forms of life (a jaw dropping rate of 20 centiMorgans per million base-pairs for those who really liked the preceding paragraph).

The impacts of this high recombination rate on honey bee health, let alone the reason it exists in the first place, are hot topics of debate. What is not debatable is that this high recombination rate is a lifesaver for scientists trying to map the key genetic features of bee biology. By chopping bee chromosomes into an aggregate of really short linked chunks from mom and dad, this means that various chunk-identifying techniques (worthy of Nobel Prizes in 1933 and 1978) can map traits to extremely small regions, even single genes. By scoring a mere 45 (haploid) drone bees that is exactly what the authors did, finding that resistance was most likely caused by mutations in a single gene that makes a protein called MBlk-1 (Mushroom body large-type Kenyon cell-specific protein-1, for you details freaks). It would be hard to predict a protein that is a better candidate for messing with mite reproduction than Mblk-1. This protein, despite the pretentious name, is actually produced in many bee cells, including the fat body which was recently shown to be the major food source of *Varroa* mites (Ramsey et al., 2019, (Ramsey et al., 2019)). More importantly, Mblk-1 is a huge enabler, that takes a change in hormone levels (i.e., the molting hormone needed to finish off a bee's transition from larva to adult) and sets in motion numerous events that help turn a pale grub into the bees we see emerging from capped cells. These changes, of course, are all happening exactly when *Varroa* mites are doing their reproductive business.

The authors suggest that the changes driven by a mutated MBLK-1 somehow throw off mite females. Perhaps mites don't get a particular smell needed to warm their ovaries? Perhaps they do not receive a vital nutritional component, or something produced by the mutated bees they are feeding on is toxic to them? Or maybe honey bee workers notice and hygienically remove bees with mutants for this protein? However, the mechanisms shake out, this trait, which arose naturally in honey bees under intense mite pressure in France, is a very attractive target for breeding. Still, in order to make this more than a good-news story for one beekeeper in Toulouse, much more needs to be resolved. Are these mutations bad for bees? Do they work for female (worker)-destined pupae? Can mites evade this suppression as they have done for human-made attacks?

I am almost naive enough to declare victory on mites, but will not do so until more of these fundamental questions are answered. It will also help to see how this protein behaves in Asian bees and against Asian *Varroa* mites. MBlk-1 levels are turned up in parasitized (drone) bees in *Apis cerana* (Ji et al., 2014), suggesting some sort of role there, or at least a response to mite attacks.

While this is not the first study to use the power of honey bee males to identify traits, nor is it the first to exploit the honey bee genome (see the latest on that in a paper by Andreas Wallberg and colleagues, "A hybrid de novo genome assembly of the honeybee, *Apis mellifera*, with chromosome-length scaffolds," (Wallberg et al., 2019)), it is exciting because it marries the two so well and it does so by tackling a terrifying threat to honey bees. No promises, but (wink, wink) keep an eye on this trait as more researchers in the near future determine how much of mite resistance it can explain.

Conlon, B.H., Aurori, A., Giurgiu, A., Kefuss, J., Dezmirean, D.S., Moritz, R.F.A. et al. (2019). A gene for resistance to the Varroa mite (Acari) in honey bee (Apis mellifera) pupae. Molecular Ecology, 28, 2958-2966.

Ramsey, S.D., Ochoa, R., Bauchan, G., Gulbronson, C., Mowery, J.D., Cohen, A. et al. (2019). Varroa destructor feeds primarily on honey bee fat body tissue and not hemolymph. Proceedings of the National Academy of Sciences of the United States of America, 116, 1792-1801.

Ji, T., Yin, L., Liu, Z., Liang, Q., Luo, Y., Shen, J. et al. (2014). Transcriptional responses in eastern honeybees (Apis cerana) infected with mites, Varroa destructor. Genetics and Molecular Research, 13, 8888-8900.

Wallberg, A., et al. (2019). "A hybrid de novo genome assembly of the honeybee, Apis mellifera, with chromosome-length scaffolds." BMC Genomics 20(275).

23

BUGS WITH BENEFITS

When it comes to microbes in honey bee colonies, the bad guys get all the press. Since Aristotle there have been countless discussions of foul scourges, blights, and paralyzing viruses that afflict bees. More benign microbes have received far less attention, if they were noticed at all. This is no longer the case. While there continue to be advances in understanding honey bee pests and pathogens, the real frontier for over a decade has involved bacteria and fungi that do not appear on bee health alerts, and are rarely mentioned in beekeeping meetings or books. Work on these 'other' bee microbes gained some footing thanks to the careful description of hundreds of microbial isolates by Dr. Martha Gilliam with the USDA in Tucson, Arizona. Dr. Gilliam was especially active in the 1970's and 1980's and, while her job also required her to study the diseases of the time, her real passion was for the many unsung microbes in honey bee nutrition. Her work (Gilliam et al., 1997) echoes that of a birder or botanist collecting samples throughout the Amazon. Her passion was to grow any and all of the bee-associated and flower-associated microbes in Petri dishes in a sterile laboratory. She isolated hundreds of types of microbes, carefully documenting where each was found, possible impacts on bees or bee products, and the overall diversity of microbial communities. Along with dissecting thousands of bees and flowers, this work used dozens of nutrient recipes and conditions to nurse microbes to grow, followed by many hours of microscopy and chemical tests aimed at putting a name on isolated microbes.

After Gilliam's work there was a bit of a lull in the study of microbes in the hive until the enigmatic Colony Collapse Disorder in late 2006. Suddenly, everything was a suspect again, since the usual suspects were not noticeably associated with all the fuss and disorder. Consequently, there was renewed interest in scouting the full microbial world for causes of bee illness. Fortunately, Professor Nancy Moran, also in Tucson at the time but at the University of Arizona, was lured into this fray. Moran had carried out many years of critical work on microbes tied to aphids and other insects and came to the bee fold with new insights and great passion. In the past decade, she and her students and colleagues have made game-changing advances in the understanding of bacteria tied to the honey bee digestive tract (Google Scholar is one place to start to see some of her work on the microbes of bees and other insects, or you can check out her Lab Page at her new home, the University of Texas, http://web.biosci.utexas.edu/moran/index.html). Collectively, they have named almost all of the key bacteria in the bee gut, tapping into bee science for names, including '*Gilliamella,*' '*Frischella,*' and '*Snodgrassella,*' as nods to previous researchers. More importantly, they have carried out fundamental work implicating bacteria in everything from nutrition to pest resistance. These bacteria have their own lives and need not be helpful for bee health, but in fact most of them seem to have a net positive effect on their bee hosts. Moran's group has also shown how human actions, from antibiotics to pesticides, can impact gut microbes and the bees that rely on them. They have also greatly energized the field of bee probiotics, an evolving topic I attempted to review here in 2017 (https://www.beeculture.com/found-in-translation-2/).

Varroa and bee with deformed wings, HFB

In a technical breakthrough just this month, Moran and colleagues used one of these resident bacteria, *Snodgrassella alvi*, as a vessel for delivering gene products that interfere with specific bee proteins, mites, and viruses. This work, led by graduate student Sean Leonard with a team of scientists from Texas, combines the power of RNA interference, a widespread mechanism for controlling genes and pathogens, with a bold attempt to tweak the cells of *Snodgrassella* to make a novel gene product (Leonard et al., 2020). RNA interference has been used in bees before, and researchers have added genetic parts to bee bacteria, but neither approach was as powerful for bee biology and health until they were married together. In brief, genetically modified *Snodgrassella* were produced in three flavors, one that knocked back a honey bee protein, one that targeted Deformed wing virus, and one that targeted snippets from 14 different genes in the *Varroa* mite. In all cases, the targets were hit and bees did better as a result. Further, the modified bacteria persisted in individual bees and social groups (all of these experiments were carried out on caged worker bees in the lab, not in colonies).

Gilliam M, Lorenz BJ, Wenner AM & Thorp RW (1997) Occurrence and distribution of Ascosphaera apis in North America: Chalkbrood in feral honey bee colonies that had been in isolation on Santa Cruz Island, California for over 110 years. http://dx.doi.org/10.1051/apido:19970601 28. doi:10.1051/apido:19970601.

University of Texas, Nancy Moran's Lab. Received at http://web.biosci.utexas.edu/moran/index.html.

Leonard SP, Powell JE, Perutka J, Geng P, Heckmann LC, Horak RD, Davies BW, Ellington AD, Barrick JE & Moran NA (2020) Engineered symbionts activate honey bee immunity and limit pathogens. Science 367: 573. doi:10.1126/science.aax9039.

24

ISLAND TIME AND RESISTANT BEES

Island populations of honey bees act as incubators for genetics. Bee genes mix and match with each other and resident microbes and pests in ways that mimic dozens of hard-won field experiments or closed breeding schemes. While it is not a given, hardy and disease-resistant bees can evolve in these isolated settings. With respect to bees that survive *Varroa* mites, islands such as Gotland between Sweden and Finland and Isla Fernando de Noronha, far off the coast of Brazil, have provided an environment for desirably resilient honey bee stock. More generally, closed breeding populations have the potential to form islands unto themselves. Perhaps the closest breeding analogy to island life is the so-called 'Black Box' strategy for producing survivor stock (recently reviewed by Tjeerd Blacquiere and colleagues, 2019, Darwinian black box selection

Hive inspection, Puerto Rico, JDE

for resistance to settled invasive *Varroa destructor* parasites in honey bees (Blacquièr et al., 2019). In 'Black Box' selection, thoughtful crosses are enacted and maintained with a focus on colony survivorship rather than any specific desired trait.

So, if islands provide such a clear path to resilient honey bees why aren't islands or isolated bee yards universally easy places to raise super bees (spoiler: they are not). In many cases, islands either 1) lack the genetic diversity and variants needed to make a leap in hardiness, 2) are subject to continuous admixture from outside populations, or 3) are sufficiently coddled by beekeepers so as to not pick up what might be costly or precarious genetic disease traits.

Among the many Caribbean Islands, Puerto Rico holds a distinct reservoir of resident genetics coupled with a laissez-faire attitude toward parasites and pathogens, a perfect mix for producing bees that are hardy in the face of disease. Much of the genetic diversity now found in Puerto Rican honey bees reflects an influx of New-World Africanized bees from the early 1990's. This influx has led to a unique bee population and a great opportunity to study how bee behaviors, positive and negative, evolve in response to a given environment. Proponents on the island also hope that their resident Puerto Rico honey bees have evolved into a stock that is desirable not only for

Dr. Ernesto Guzman, conducting frame inspections, JDE

Puerto Rico but perhaps for other regions with Africanized bees.

Professor Tugrul Giray and his students and collaborators at the University of Puerto Rico are actively studying the behaviors and genetics of Puerto Rican bees. Their data suggest that two things have occurred since the arrival of AHBs to the island. First, honey bee populations there tolerate *Varroa* mite parasites quite well with no help from beekeepers. Beekeeping occurs across the island in dry forests and plains as well as mountains and rain forests. While Puerto Rican bees are not subjected to long winters, they do maintain seasonality mediated by patterns of rainfall, and they thrive across the various island ecosystems. Hurricane Maria, which devastated plants, bees, and people in 2017, was followed by a resurgence of all three. It is tempting to think that an influx of AHB genes has helped Puerto Rican populations get ahead of mites.

Second, Puerto Rican honey bee populations are in the gentler range of AHB, in terms of defensive behaviors. Scientists there suggest this is a combination of an adaptation to seasonal forage availability, restricted gene flow, selective pressures from humans in the island and lack of other predators (e.g., Bert Rivera-Marchand and colleagues, 2012 (Rivera-Marchand et al., 2012). Release from predators is a common trait of islands, leading to the evolution of flightless birds and 'tame' animals, at least until humans intervened. Whatever the mechanisms, the bees of Puerto Rico, on average, have gone from typical AHB feistiness early on to a more 'simpatico' or gentle state, leading Professor Giray and his colleagues to label them gentle-AHBs, or gAHBs. They have done this rapidly, while retaining an impressive feistiness towards mites and other disease. Given the unique history of these bees, along with a desire to tease apart complex behaviors, gAHBs have been subjected to state-of-the-art genomic analyses. Arian Avalos and colleagues (Avalos et al., 2017) worked up entire genome sequences for 30 Puerto Rican gAHBs, 30 New-World European bees and 30 New-World Africanized bees from the mainland (think '23-and-Bee'). Along with showing that Puerto Rican bees hold a signature in their genes that is unique from other populations, these researchers identified genome regions that unite the gAHB population with EHBs, possibly indicating the very traits that keep them calm. While research on the Puerto Rican honey bee population continues, the results thus far indicate that gAHBs have a mix of meaningful genetic traits from both EHB and AHB parent populations. If Puerto Rican beekeepers continue to focus on gentle traits and self-sufficiency against disease, this population seems likely to continue its path toward sustainability.

I am always happy to inflict my flawed Spanish on native speakers and was thrilled to be invited to a conference regarding the bees of Puerto Rico and their environment, sponsored by the National Science Foundation, University of Puerto Rico, the Puerto Rico Science, Technology and Research Trust, and Florida International University (prhb.cs.fiu.edu). Attended by 100 beekeepers from across the island and dozens of students and scientists, this forum provided a chance for updates on science and broad discussions about bees, disease, and beekeeping in Puerto Rico and elsewhere. It also provided a full day in the Puerto Rican countryside with bees and beekeepers.

The author inspecting hives in Puerto Rico, JDE

While there is justifiable excitement over the unique blend of genes found in Puerto Rican honey bees, decisions to transport these bees to the mainland U.S. or elsewhere will be made with caution, knowing how hard it is to rewind the clock if, for any reason, things go badly. Soon, a team of individuals will begin to develop a logical framework for assessing Puerto Rican bees and their associates for their good, bad, and undetermined traits. These assessments will be carried out by specialists and eventually used to help shape a logical framework for the regulators who will have the final say.

In the meantime, there is no doubt that the bees and beekeepers of Puerto Rico are resilient and unique in many ways. Moreover, the excellent studies of behaviors, colony traits, and underlying genetics that have been carried out thus far provide a framework for exploring the honey bees of Puerto Rico as well as other island populations, and indeed honey bee subpopulations more generally.

Blacquièr, T., Boot, W., Calis, J., Moro, A., Neumann, P. & Panziera, D. (2019). Darwinian black box selection for resistance to settled invasive Varroa destructor parasites in honey bees. Biological Invasions, 21, 2519–2528.

Rivera-Marchand, B., Oskay, D. & Giray, T. (2012). Gentle Africanized bees on an oceanic island. Evolutionary Applications, 5, 746-756.

Avalos, A., Pan, H., Li, C., Acevedo-Gonzalez, J.P., Rendon, G., Fields, C.J. et al. (2017). A soft selective sweep during rapid evolution of gentle behaviour in an Africanized honeybee. Nature Communications, 8. doi:10.1038/s41467-017-01800-0.

25

BEES, BEENOMES, AND BENEFITS FROM SCIENCE

Close your eyes and imagine a realm where you are seeing features of biology for the very first time, allowing you to piece together diverse insights you would never have dreamed of just moments before seeing them…insights and connections that change decades of scientific thinking. No, this is not a stroll through a tropical rain forest (though that sounds nice, too), but the first look scientists get when they sequence, assemble, and investigate an organism's genome. Honey bee's faced this scrutiny a decade ago and the results continue to provide insights into not just honey bee biology but much of life as we know it. I described the methods and some driving forces for this effort in Bee World (Evans, 2005), but the key paper describing the honey bee genome went public in the journal Nature in 2006, (Weinstock et al., 2006). Scientists have cited this paper over

Bee contemplating its genome, PG

1000 times in the past decade and the bee genome has helped diverse scientific fields including animal behavior, insect development, and the study of how plants and animals follow a regular daily 'clock.' On the side of bee health, the genome highlighted the fortress of bee sociality alongside potential weaknesses in bee defenses toward disease and pesticides (redundant evidence to some).

Recent efforts to interrogate the honey bee genome have sped the development and maintenance of favored breeds and have had an impact on studies of bee stress, disease, and behavior. Some of this work was on display at the 2018 "Biology and Genomics of Social Insects" workshop at the historic Cold Spring Harbor Laboratories (CSHL) in New York. This meeting gave team members of the 'Beeomics' Consortium (http://www.beeomics.ca/) a chance to share results from their efforts to use genomic insights to identify desirable disease, overwintering, and behavioral traits. In preliminary work with markers developed early in their sequencing efforts, they showed a significant economic benefit gained when breeders used genome-based markers as part of their decision process (Miriam Bixby and colleagues at University of British Columbia and elsewhere, in the Journal of Economic Entomology, (Bixby et al., 2017)). Marker-based efforts mesh well with classical Varroa resistance breeding. Similarly, the Beeomics group is now finalizing a promising new genetic screen for Africanized bees, a key regulatory goal.

Two innovative genome studies described at the CSHL meeting involved measuring the distinctive traits of 'winter' honey bees. First, Harshil Patel and members of Amro Zayed's group at York University, Toronto, introduced a collaborative project that has sequenced the genomes of 1000 honey bee samples from colonies across North America These colonies were otherwise unbothered, giving colony metrics for overwinter survival that can now be matched with specific sets of genes. Winter bees have been scrutinized in many ways, including work by another Canadian, Gard Otis, some years ago (Matilla et al., 2001) but the source and traits of quality winter bees remain hot questions. To that end, Tomas Erban and colleagues in the Czech Republic used a genome-enabled approach to show that winter bees carried an abundance of vitellogenin (a key protein in bees involved with everything from reproduction to resilience) and other proteins linked with nurse bees and bees having excellent nutrition (Erban et al., 2013). In the most ambitious effort yet, work described at CSHL by Mehmet Doke, Tugrul Giray, and Christina

Grozinger (Penn State Univ. and Univ. Puerto Rico) followed the expression levels of all active genes in the honey bee body as bees entered winter. They compared these active genes with summer foragers and nurse bees. The results led them to propose that winter bees are in some ways a third worker form, one that takes the best of nurse bees (an over-active fat body in terms of proteins linked with on-demand energy) and foragers (an over-active set of wing muscle proteins, which they suggest make these winter bees better able to keep the cluster warm in winter); neat stuff, and now testable in the field. Specifically, these markers can be used to vet the different ways used by beekeepers to prepare their bees for winter, perhaps changing established thinking on what we do to set up bees for their greatest challenge at the colony level.

As the usual disclaimer, gene-based approaches will not prove useful for all problems facing bees, and many beekeepers can and will get along just fine without these insights. Still, it can't be denied that this is a really exciting field of science, if only because it helps us better understand the strengths, weaknesses and novelties of an amazing insect.

Evans, J.D. (2005). Beenome-mania: How will the honey bee genome project help beekeepers? Bee World, 86, 25-26. Received at: https://www.researchgate.net/publication/298550303_Beenome-mania_how_will_the_honey_bee_genome_project_help_beekeepers.

Weinstock, G.M., Robinson, G.E., Gibbs, R.A., Worley, K.C., Evans, J.D., Maleszka, R. et al. (2006). Insights into social insects from the genome of the honeybee Apis mellifera. Nature, 443, 931-949. https://www.nature.com/articles/nature05260.

Beeomics. Sustaining and Securing Canada's Honeybees. Received at http://www.beeomics.ca/.

Bixby, M., Baylis, K., Hoover, S.E., Currie, R.W., Melathopoulos, A.P., Pernal, S.F. et al. (2017). A Bio-Economic Case Study of Canadian Honey Bee (Hymenoptera: Apidae) Colonies: Marker-Assisted Selection (MAS) in Queen Breeding Affects Beekeeper Profits. Journal of economic entomology, 110, 816-825. doi: 10.1093/jee/tox077.

Matilla, H.R., Harris, J.L. & Otis, G.W. (2001). Timing of production of winter bees in honey bee (Apis mellifera) colonies. Insectes Sociaux, 48, 88-93.https://link.springer.com/article/10.1007/PL00001764.

Erban, T., et al. (2013). "Two-dimensional proteomic analysis of honeybee, Apis mellifera, winter worker hemolymph." Apidologie 44(4): 404-418.

26

SOCIAL STATUS AND THE SINGLE BEE

After 20 years in a honey bee research lab, I am biased towards this hero of the air, agriculture, and environment. That being said, it helps to reflect on the diversity of bees worldwide and the many ways they persist. The vast majority of the world's 20,000+ bee species live as 'single moms,' Here, female bees gather pollen and nectar over hundreds of trips, packing these resources into tunnels or other cavities and laying a single egg atop each 'pollen ball.' They then seal things up and die without meeting any of their offspring. Additional bee species nest communally (think cliff swallows, or bats in a cave). Very few bee species form truly social colonies. Of the social bees, even fewer form huge, highly interactive societies like honey bees. Nevertheless, honey bees show their bee roots in many ways, from their flight muscles to their needs for pollen and nectar and their general sensitivity to chemicals. Similarly, solitary bees show many traits found in their social relatives, from repeated risky

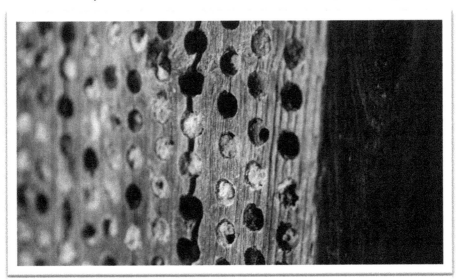

Communal mason bees, HFB

trips to collect food they will never eat to an ability to choose which eggs will develop into male versus female offspring.

A few species teeter on the edge of sociality, with female offspring sometimes staying with their mothers to help raise new females and sometimes setting off to mate and try their luck alone. This switch can reflect genetics or environmental conditions such as the time of year or the availability of resources. Sweat bees provide some of the best-known and best-studied examples of 'polymorphic' sociality. The late Cornell bee scientist George Eickwort worked painstakingly to describe the biology of the widespread North American sweat bee *Halictus rubicundus*. This generally social species shows a solitary lifestyle in regions with a short growing season, such as high in the Colorado Rocky Mountains (Eickwort et al., 1996). While Eickwort and others (including his children) investigated burrows of *Halictus* with spoons and tiny flags, Cécile Plateaux-Quénu was studying a similar sweat bee, *Lasioglossum albipes*, in France. Populations of this species, found in locations with longer, warmer, summers tend to be social. Here, female foundresses produce a small number of workers (all female) and males each spring. The workers then stick around to collect food for a second generation of female sisters and males. This second-generation mates, whereupon males die (another similarity with honey bees) and females dig in for winter before emerging and starting new nests the following spring. Social habits seem to have a genetic component of some sort in this species, in that females from the social and solitary forms who are collected and placed in identical environments are faithful to their original social behaviors (Plateaux-Quénu et al., 2000). The ancestor of *L. albipes* was consistently social, so this is really a story of a social bee that sometimes opts for a single-mom life, with eggs left to hatch on their own, and sometimes sets the stage for a generation of daughters to stay around and invest in the family business.

The fascinating biology of *L. albipes* might have faded from view had this species not attracted the attention of Princeton biologist Sarah Kocher (https://www.kocherlab.com/). Kocher carried out years of field work to further describe the social habits of these bees, in time identifying more solitary and social populations across France. With students and colleagues, she then began an experimental dissection of what makes members of this species opt for producing largely sterile workers at times, and only future egg-layers at other times. These efforts are now shedding light on social traits shared with honey bees and other insects, and perhaps even ourselves. One study, published in 2017 (Wittwer et al., 2017), showed that the social forms of *L. albipes* have consistently more sensilla, organs on their antennae known to be involved with picking up chemical cues, when compared to solitary populations. Looking across 36 sweat bee species, solitary species that were at the tips of social lineages tended to carry fewer of these smell-sensing sensilla. Sensilla are key for picking up on subtle social cues in bees (and ants!) and the authors propose that the apparent loss of these organs in solitary species is analogous to the decrease in eye size found in fish or other species that inhabit caves. Put another way, these organs are likely costly for bees, and if they are strictly involved with social behaviors, then solitary bees do better without them.

Not content to measure behaviors or antennal organs alone, Kocher and crew sequenced the genome of this bee (at the time it was one of the first bee species to have its genome exposed to scientists) and used this blueprint of bee proteins to further explore social and solitary differences. Next, they published a stunning paper that gives a comparative look at the genomes of 143 bees from both social and solitary populations of *L. albipes* (Kocher et al., 2018). The study confirms that social life is optional in this species, with bees from specific regions tending to be social and from other regions tending to be solitary. Social bees shared common traits in their genomes, giving the best view yet of specific proteins and protein variations that might help drive sociality. The most striking variation was found near the gene encoding the protein syntaxin 1a. This protein is known to be over-expressed with respect to social behavior in other species and is even, ta-da!, known to be involved in the learning of smells by honey bees. Sure enough, social sweat bees over-expressed this protein relative to solitary ones, consistent with a direct role in social communication. Other genes linked with sweat bee sociality also showed connections to the behavior of honey bees, and even humans, a topic that will take more time to explore. It remains to be seen how these results will improve our understanding of how honey bee societies work, or be used to improve beekeeping. Still, comparative studies across bees that are social, solitary, or on the fence, are clearly poised to inform what it takes to make highly social insects like honey bees succeed.

Eickwort, G.C., Eickwort, J.M., Gordon, J. & Eickwort, M.A. (1996). Solitary Behavior in a High-Altitude Population of the Social Sweat Bee Halictus rubicundus (Hymenoptera: Halictidae). Behavioral Ecology and Sociobiology, 38, 227-233. https://www.jstor.org/stable/4601196.

Plateaux-Quénu, C., Plateaux, L. & Packer, L. (2000). Population-typical behaviours are retained when eusocial and non-eusocial forms of Evylaeus albipes (F.) (Hymenoptera, Halictidae) are reared simultaneously in the laboratory. Insectes Sociaux, 47, 263-270. https://link.springer.com/article/10.1007/PL00001713.

The Evolution of Social Behavior. Kocher Lab at Princeton University. Received at https://www.kocherlab.com/.

Wittwer, B., Hefetz, A., Simon, T., Murphy, L., Elgar, M., Peirce, N. et al. (2017). Solitary bees reduce investment in communication compared with their social relatives. PNAS, 114, 6569-6574. http://www.pnas.org/content/114/25/6569.

Kocher, S.D., Mallarino, R., Rubin, B.E.R., Yu, D.W., Hoekstra, H.E. & Pierce, N.E. (2018). The genetic basis of a social polymorphism in halictid bees. Nature Communications, 9. https://doi.org/10.1038/s41467-018-06824-8.

27

WEATHER, YOUR BEES LIVE OR DIE

Honey bee colonies, along with humans and the rest of life on Earth, are strongly impacted by the weather. As a species, *Apis mellifera* has succeeded incredibly well from the tropics to the colder regions of Europe and Asia. With help from their human keepers, honey bees now live across most of the globe, surviving drought conditions, intense rain, and winters that have them sheltering for three fourths of the year. Still, bees and beekeepers are not completely immune from climate impacts, both current and forecasted. Understanding the impacts of climate on forage and bee needs is vital for thoughtful beekeepers as they make decisions about bee movement, provisioning, and parasite control.

A recent study shows the value in exploiting high-resolution weather data in order to predict climate impacts on honey bee health and overwinter survival. Thanks to our economic and emotional obsession with the weather, satellites and ground weather stations collect continual data at a fine scale throughout the world. Matthew Switanek and colleagues from the University of Graz in Austria used creative methods to connect weather data to bee colony fates in their recent paper "Modeling seasonal effects of temperature and precipitation on honey bee winter mortality in a temperate climate" (Switanek et al., 2017). The scientists involved include two climate experts as well as Karl Crailsheim and Robert Brodschneider, leading bee researchers from Graz.

The authors used computational tools and weather station data to infer temperature, rainfall, precipitation, wind speed, and solar radiation on a 5-kilometer (km) x 5 km grid across the entire country, a scale that matches honey bee foraging distances. They merged five years of weather reports with colony overwintering loss rates for the same time span and let the data tell them which weather variables aligned best with individual colony loss rates. Thanks to the mountainous nature of Austria they found great variation in all of these traits both across months and years and across apiaries in the country. They also benefitted from an open survey of beekeepers and precise colony locations for >100,000 colonies across the five-year study. The most important predictors for colony loss were temperature and precipitation; higher temperatures and low rainfall in almost every month were both tied to future colony losses. An exception that lends credence to their methods was February, for which extreme cold snaps reversed the temperature effect and had a negative relationship with colony survival. It is important to note that the computer models used by the authors separated each variable, even those that are naturally correlated (i.e., sunny, warmer days are tied to less rainfall). Again, this gives more confidence that they are identifying causal factors in colony survivorship.

In the end, they used their weather grid to 'predict forward' weather impacts, using colonies not included in their models. Along with producing a spectacularly colorful graph (this essay), this effort did quite well in forecasting overwinter mortality for Austrian beehives. The results do not overwhelm the many additional factors that impact colony health, from parasites and pesticides to nutrition, but rather show how weather

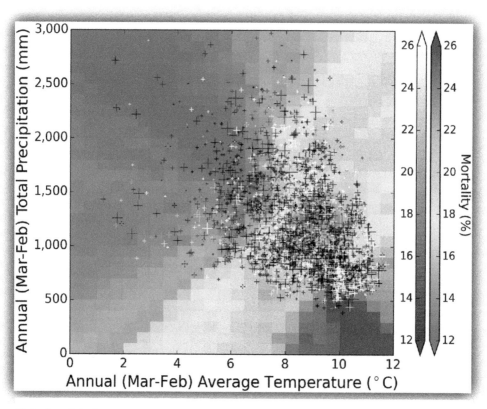

Relationship between temperature, precipitation, and bee health, RB

interacts with these factors. Indeed, good beekeepers, given colony checkups and their own climate sense, can and do adjust to weather all the time. Weather and seasonality are no-brainers for mite treatments, colony transport, and nutrition for example. Nevertheless, these results indicate that weather had a strong and often quite local background impact on winter losses on top of beekeeping decisions and variable external pressures.

On a practical level, beekeepers might be able to use this month's or last month's rain regime as a predictor of forage long before those plants start flowering. On a larger scale, honey producers and agricultural statisticians could use, and likely are using, weather patterns worldwide to infer colony health and productivity. For both efforts, the computational tools used by these authors might help decouple the weather from the many other factors beekeepers and producers juggle when making their decisions.

If you are curious to see for yourself the precision of weather data and even apply it to your past notes regarding bee survival, treatment and feeding, there are excellent and freely available resources. As a start, plug your hometown or apiary zip code into the web at https://www.wunderground.com/history/. If you choose a random date from the past, you will see all the basic weather information you might care for. If you are a multi-generation beekeeper, or especially lucky, you can compare your colony notes to daily or monthly weather records from the 1930's onward for some sites in the United States. International locations listed in this site seem to be limited to the past 20 years, still a good time frame for those curious to see how the weather matches their own beekeeping successes and losses. Should you wish to carry out your own analyses of long-term weather effects on local honey bee colonies, you can enter extended dates. With little effort, I was able to get a spreadsheet file showing daily numbers for all the basic weather parameters in College Park, Maryland, from 1948 to the present time, fun!

In Essay 18, I discussed ways in which bee genes become matched to their environment, a hot topic in beekeeping and bee breeding. I was privileged to watch a fascinating talk by Erin MacGregor–Forbes at the 2017 North American Beekeeping Conference. With support from the federally funded Sustainable Agriculture Research and Education program (www.SARE.org), she conducted a timely experiment to determine colony fates when honey bee packages headed by southerly queens were re-queened with queens reared from populations closer to her Maine study site (http://mysare.sare.org/sare_project/fne12-756/). Indeed, this manipulation had a strong positive effect. Time will tell whether these striking results reflect climate-adapted genes, queen quality, or perhaps the benefits of re-queening mid-season, but the potential is great.

Switanek, M., Crailsheim, K., Truhetz, H. & Brodschneider, R. (2017). Modelling seasonal effects of temperature and precipitation on honey bee winter mortality in a temperate climate. Science of the Total Environment, 579, 1581-1587. http://dx.doi.org/10.1016/j.scitotenv.2016.11.178.

Weather Underground, Historical Weather. Received from https://www.wunderground.com/history/.

Sustainable Agriculture Research and Education. Grants and Education to Advance Innovations in Sustainable Agriculture. Received from www.SARE.org.

Sustainable Agriculture Research and Education. Project Overview: A comparison of strength and survivability of honeybee colonies started with conventional versus northern re-queened packages. Received from http://mysare.sare.org/sare_project/fne12-756.

28

DELAYED MORTIFICATION

Thousands of studies have shown the impacts of disease, chemical stress, and poor nutrition on honey bee queen, worker, and drone longevity. Acute insults from any of these routes can kill bees in days. What has been harder to measure is how challenges affect bee behavior and productivity across entire lifespans and in the context of the colony as a whole. Several recent studies provide insights into colony-level impacts of disease and stress, providing new avenues for measuring these factors in a way that is important for bees and beekeepers.

The first challenge has been to translate 'sublethal' events into traits that individual bees use to provision, protect, and maintain their colonies.

Fungicide treatment in almond orchard, HFB

Honey bees are notoriously smart for insects, so an appropriate target for such studies is to measure how individual bees learn tasks. In repeated studies, honey bee workers learn relevant tasks (such as the ability to associate a smell with a food reward) at a lower rate after exposure to disease agents and certain chemical stresses. Zhiguo Li and colleagues describe possible mechanisms behind these learning deficits in their paper "Brain transcriptome of honey bees (*Apis mellifera*) exhibiting impaired olfactory learning induced by a sublethal dose of imidacloprid" (Li et al., 2019). After low-level pesticide exposure, worker honey bees showed no increased mortality but did poorly on a test of their abilities to associate a smell (lemon) with a food reward. The brains of treated bees were then analyzed for gene activity. Bees under chemical stress showed reduced levels for genes arguably involved with sensory perception and learning, among others. The behavioral results in this study are similar to those found in worker honey bees following infection with Deformed wing virus (Javaid Iqbal and Uli Mueller, "Virus infection causes specific learning deficits in honeybee foragers," (Iqbal & Mueller, 2007)). In both studies exposed or infected worker bees did not show obvious symptoms and lived to a typical age, in some cases even longer than expected. The presumed impact was on their legacy of providing for the colony as a whole.

As we tackle long-term effects on bees, there is a need to resolve how good and bad events alike interact to affect bee health and productivity. A striking recent example of such interactions involves the simultaneous exposure of honey bees to two agrochemicals. In a careful pairwise study, Andrea Wade and colleagues found a 2000-fold increase in adverse impacts when a particular fungicide and insecticide were presented together in "Combined toxicity of insecticides and fungicides applied to California almond orchards to honey bee larvae and adults" (Wade et al., 2019). This information was immediately used by a sympathetic orchard industry to alert growers. Nature being nature, these synergistic interactions seem rare. This is good news for regulators and industries that wish to reduce impacts, but challenging in that the reasons behind these synergisms are still mysterious. In the meantime, identifying the myriad of co-occurring insults picked up by honey bees is a big challenge, yet that is exactly what is needed to direct future research into possible synergists.

Interactions often must be measured across long time periods and at the level of colonies. On the plus side, honey bee colonies provide a buffer of sorts against disease and abiotic stress. Whether because of an ability of colonies to shift behaviors or resources, genetic diversity that allows at least a fraction of the colony to escape threats, or simply a huge 'body' that is harder to perturb, honey bee colonies can endure stresses that are consistently lethal for solitary pollinators or those with smaller colonies. This 'superorganism' benefit likely prevents honey bee losses from being even higher than those observed (read a review by Lars Straub and colleagues, "Superorganism resilience: eusociality and susceptibility of ecosystem service providing insects to stressors," (Straub et al., 2015)). On the down side, honey bee colonies are a target of numerous parasites and pathogens. Once they get established, these agents can be additive over the season or lifetime of colonies, increasing risks to both bees and entire colonies months later.

Richard Odemer and colleagues in Germany describe the most recent attempt to measure the combined and individual impacts of chemicals and disease on bees in the field. Their work, "Sublethal effects of clothianidin and *Nosema* spp. on the longevity and foraging activity of free flying honey bees" (Odemer et al., 2018), reflects an ambitious study to expose field colonies to field-relevant levels of a common neonicitinoid pesticide, measure residues in bee stores and bees themselves, and then see how these exposures impact bees in the presence or absence of an induced *Nosema* infection. Exposure to *Nosema apis* led to higher bee mortality rates but low-level exposure to clothianidin did not. Surprisingly, combined exposure to both threats showed no synergism, in contrast to several laboratory studies, perhaps reflecting the resilience of the superorganism. Also, in *Ecotoxicology*, Reinhold Seide and colleagues describe a similar project aimed at identifying the impacts of clothianidin on disease levels (Reinhold & Seidel, 2018). In a study of 24 colonies that spanned nearly one full year, colonies given an initial dose of 200 ppb clothianidin died within ca. 50 days, while those receiving lower doses survived at the same rate as the controls and maintained similar worker numbers. They observed trends toward higher mite levels in colonies exposed to low levels of clothianidin (10 and 50 ppb in nectar at the start of the experiment) but no significant differences for levels of mites, *Nosema*, or viruses. Both *Ecotoxicology* studies were strengthened by the careful measurement of pesticide levels in exposed and control hives, giving insights into the fates of field-collected nectar contaminated by pesticides. One caveat to the work with low-level exposure to pesticides was that the observed residues in bees were even lower, in part because spiked nectar was further diluted by outside sources. As with most field studies, the results were also weakened by small sample sizes. Quantifying subtle differences in disease loads and survivorship requires screening on the order of 50 or more colonies, given the many unknown factors and chance events that cause bee colonies to differ in important traits.

As challenging as it might be to expose worker bees to multiple stresses, it is even harder to design experiments that measure how challenges on one life stage or caste (queen, worker, or drone) affect later ones. Christina Mogren and colleagues faced this challenge by measuring the impacts of larval nutrition on the abilities of worker bees to survive adult challenges ("Larval pollen stress increases adult susceptibility to clothianidin in honey bees," (Mogren et al., 2019)). To do this, they deployed pollen traps to rob pollen from ten nucleus colonies and then fed a fraction of this pollen to ten additional nucleus colonies in the form of patties. After four sessions of pollen distribution, frames were removed from both colony types and worker bees were allowed to finish development and emerge in incubators. Newly emerged bees were reared in cups on a diet of sucrose syrup spiked with the insecticide clothianidin at concentrations of 10, 40, 200 and 400 parts-per-billion. Bees from pollen-starved colonies did especially badly when they subsisted on syrup with 40 and 200 ppb of the insecticide. While pollen-starved bees also fared more poorly than pollen-supplemented bees when maintained on non-contaminated sugar water, the experiments did suggest that both physiological changes and mortality in bees exposed to chemicals were affected by larval nutrition.

A large fraction of honey bee research is now focused on the challenges of determining the key impacts of disease and stress on the colony level. Hopefully, this work will lead to additional insights and recommendations for maintaining colony health in the face of these challenges. Honey bees thrive in the face of a range of environmental challenges, in part because they maintain a sisterhood of thousands of workers, but even honey bees will need more help to persist in the face of these challenges. Complex field research projects, and over-arching analyses across many such studies, are helping to identify the main factors that bug honey bees, the first step in managing these threats.

Li, Z., Yu, T., Chen, Y., Heerman, M., He, J., Huang, J. et al. (2019). Brain transcriptome of honey bees (Apis mellifera) exhibiting impaired olfactory learning induced by a sublethal dose of imidacloprid. Pesticide Biochemistry and Physiology.

Iqbal, J. & Mueller, U. (2007). Virus infection causes specific learning deficits in honeybee foragers. Proceedings of the Royal Society B: Biological Sciences, 274, 1517-1521.

Wade, A., Lin, C.H., Kurkul, C., Regan, E.R. & Johnson, R.M. (2019). Combined toxicity of insecticides and fungicides applied to California almond orchards to honey bee larvae and adults. Insects, 10.

Straub, L., Williams, G.R., Pettis, J., Fries, I. & Neumann, P. (2015). Superorganism resilience: Eusociality and susceptibility of ecosystem service providing insects to stressors. Current Opinion in Insect Science, 12, 109-112.

Odemer, R., Nilles, L., Linder, N. & Rosenkranz, P. (2018). Sublethal effects of clothianidin and Nosema spp. on the longevity and foraging activity of free flying honey bees. Ecotoxicology, 27, 527-538.

Reinhold and Seidel (2018). "A long-term field study on the effects of dietary exposure of clothianidin to varroosis-weakened honey bee colonies." Ecotoxicology 27(7): 772-783.

Mogren, C.L., Danka, R.G. & Healy, K.B. (2019). Larval pollen stress increases adult susceptibility to clothianidin in honey bees. Insects, 10.

29

OVER IN WINTER

I don't want to dwell on loss, but there are increasingly good studies recently that take the long view on what causes colonies to fail between fall and spring. In the U.S. and many temperate countries, overwinter losses remain stubbornly high. There must be an explanation there, or perhaps several. As the lab fund-holder, it pains me that even here we have to shell out for packages every spring (and I LIKE the guy we get packages from).

As with other times of year, unwarranted winter losses (e.g., colonies that went into winter with sufficient stores and numbers to have a chance of surviving) likely reflect a mix of stresses. Marco Beyer and colleagues describe their search for the most important of those stresses in an upcoming paper from *Research in Veterinary Science* (Beyer et al., 2018). By diligently following hives via repeated surveys across four years (1364 'hive-years' in total) they were able to tease apart trends related to both mite treatment strategies and climate. Interestingly, both this study and the Austrian study by Matthew Switanek and colleagues I highlighted in Essay 27 (Switanek et al., 2017), found that that warm winters correlate with higher winter losses. Whether that is due to maladaptive choices by queens or workers, a failure to re-cluster in cold snaps, or a more favorable environment for pests and parasites is unclear. Whatever the proximate cause, the impacts of warm winters were observed fairly suddenly. Both studies found that warm December and January temperatures were a solid predictor of colony death the very same winter. In addition, both studies showed believable trends between inopportune rainfall and colony losses (e.g., rainy July's led to higher colony losses while rainy Septembers favored colony overwinter survival).

Lest this seem like a slam-dunk for the effects of climate on overwintering success, the authors found that mite treatment regimens by surveyed beekeepers had an equally large impact on overwinter losses. In particular, they identified a strong difference in overwintering success based on the strength of formic acid used to treat colonies. Colonies receiving a 60% formic acid treatment tended to fare poorly in this survey while those given a greater mite shock (85% formic acid) survived much better. Unfortunately, they do not couple this observation with data on actual mite loads, and they make clear that the weaker formic treatments might have been applied differently, but on the whole the differences were striking. The next strongest predictor of high colony survivorship came for beekeepers using oxalic acid midwinter, although this by itself was not enough to significantly decrease losses.

In another fresh paper from the *Journal of Apicultural Research*, Philip Brown and colleagues attempt to determine drivers of colony loss in New Zealand (Brown et al., 2018). Thanks perhaps to the allure of pricey honey, New Zealand colony levels have tripled in the past ten years and are six times higher than they were in the 1940's, a sharp

Honey bee colony facing winter stress, JDE

contrast with the U.S. where bee numbers have declined substantially in that same time frame. Another reason for the great interest in beekeeping there might be that New Zealanders have not fully met the heartache of loss, with average loss rates near ten percent (sigh). *Varroa* and *Nosema ceranae* are features on both of the major New Zealand islands but seem not to have the same impacts yet. In fact, on the side of arthropod pests, key threats identified in the study included wasps and an aphid! (ok, not the world's first bee-attacking aphid but the giant willow aphid which, true to its name, damages an important pollen source).

The author with midwinter colonies, JDE

Back to the Northern hemisphere, expanding data coming from the Bee Informed Partnership continues to show a connection between mite control or the lack thereof with both overwinter and yearly losses. The always-improving survey app at https://bip2.beeinformed.org/survey/ takes the work out of poring over their growing treasure trove of data.

As in past years, mite treatments, and amitraz in particular, are tied with better fates. Interestingly, only 25/234 respondents who were commercial beekeepers did NOT use amitraz in 2016 and 2017 combined, but for those 25, loss rates were equal to those using amitraz. All other classes of beekeepers showed significantly higher losses if they skipped using amitraz. Solid surveys from the equally stricken (> 30% losses) Netherlands also show mites to be the greatest factor, with pesticides in honey and the presence of canola pollen skating into the silver and bronze spots. Romée van der Zee (e.g., (Van Der Zee et al., 2015)) continues to rally good survey numbers in that country, as part of the global Colony Loss Network (www.coloss.org).

So, how can these survey results help you? For one, they are only strong when everyone takes part, so be sure to reply (in the U.S.) to both the BIP survey and the USDA's National Agricultural Statistics Service survey on colony health. And keep perusing their results to compare with your own experience. The results are free and continue to provide good insights into managing bees. You can't change the weather (as individuals), but you can react to it smartly and there are other risks that are more easily addressed.

Beyer, M., Junk, J., Eickermann, M., Clermont, A., Kraus, F., Georges, C. et al. (2018). Winter honey bee colony losses, Varroa destructor control strategies, and the role of weather conditions: Results from a survey among beekeepers. Research in Veterinary Science, 118, 52-60. https://doi.org/10.1016/j.rvsc.2018.01.012.

Switanek, M., Crailsheim, K., Truhetz, H. & Brodschneider, R. (2017). Modelling seasonal effects of temperature and precipitation on honey bee winter mortality in a temperate climate. Science of the Total Environment, 579, 1581-1587. http://dx.doi.org/10.1016/j.scitotenv.2016.11.178.

Brown, P., et al. (2018). "Winter 2016 honey bee colony losses in New Zealand." Journal of Apicultural Research: 1-14.

Bee Informed Partnership. National Management Survey. Received at https://bip2.beeinformed.org/survey/.

Van Der Zee, R., Gray, A., Pisa, L. & De Rijk, T. (2015). An observational study of honey bee colony winter losses and their association with Varroa destructor, neonicotinoids and other risk factors. PLoS ONE, 10. e0131611. doi:10.1371/journal.pone.0131611.

COLOSS Honeybee Research Association. Received at www.coloss.org.

SNOWBIRDS, SNOW, AND SUPPLEMENTS SHED LIGHT ON OVERWINTERING SUCCESS

November is too late for most of us to plan for this winter, but insights continue to arrive from studies that connect summer and fall planning to more and stronger hives in the spring. Previously, I've highlighted papers that relied on beekeeper survey responses and various external measurements (weather, local land use) to model colony losses and infer causes (Essay 29). More such surveys are now out and the scale and details of these surveys are only getting better.

Researchers are also describing ambitious colony, and even habitat, management experiments that strengthen arguments for

Early spring colonies in almonds, HFB

manage bees for overwintering success, and how to find the best sites for forage and 'storage' during dearths. Vincent Ricigliano and colleagues with the USDA-ARS in Tucson, Arizona have provided an important study that documents the impacts of queen replacement and food on winter success, "Honey bees overwintering in a Southern climate: longitudinal effects of nutrition and queen age on colony-level molecular physiology and performance" (Ricigliano et al., 2018). Importantly, this work is from the Imperial Valley of California. Near the Arizona border, this site is far more southerly than prior studies. In November, average high temperatures in nearby Yuma, Arizona, are 77 degrees Fahrenheit and almost certainly sunny…but such is winter there. Many beekeepers find this climate suits them for feeding bees and building numbers prior to almond pollination.

In the Ricigliano study, starting the season with replacement queens of matched Italian stock had a significant impact on brood production some months later, increasing production by around 25% averaged across all colonies. In addition, adding a mix of Michigan wildflower and California almond pollen to a pollen substitute tended to increase brood production, but not significantly given other factors from the field. Nevertheless, nine colony pairings that were scored in November and January (and split by site and queen replacement) showed higher brood production in the 'real-pollen' set while three such pairings showed lower production when pollen was added. Curiously, the three colony measurements that bucked the trend were at just one of three apiaries. In other words, at two of three sites, 'true' pollen increased brood production for each of the sampling points. As in every field study to date (or beekeeping operation, for that matter) there was far more variation colony-to-colony than could be explained by the researchers' manipulations. Two months after colonies were equalized, colonies in each set differed from each other by over 50% in either direction in terms of brood area and this 'noise' in the system persisted until the end of the trials. A million-dollar question remains why do equalized colonies of the same stock, same management, and same location take such different paths in a couple months?

One way to solve this puzzle is to query the bees themselves to measure the physical state of individual bees. The tool of choice for this currently is to assess the activity levels of key honey bee genes linked to disease resistance, stress and overall robustness. In the Ricigliano study, those genes included immune factors (which increased upon entering 'winter') as well as a family of established markers for honey bee adult development and nutritional status (e.g., vitellogenin, whose many properties are summarized by Miguel Corona and colleagues at (Corona et al., 2007)). Expression levels of several immune genes and vitellogenin increased as the colonies entered November and December, consistent with a 'winter bee' response, however benign that winter turned out to be. The authors also describe levels of additional genes in the vitellogenin 'family', one of which (vg-like 'a') was actually a better predictor of seasonal status than vitellogenin itself. This result alone might help improve the ability of researchers and beekeepers to track the health status of colonies in different conditions and management regimes, perhaps clearing up some of the colony noise.

Another way to separate truth from noise is by brute force. When thousands upon thousands of colonies are tracked, individual factors that help colonies fail or prosper should emerge. I have reviewed before important efforts by both the USDA National Agricultural Statistics Service (https://www.nass.usda.gov/Surveys/Guide_to_NASS_Surveys/Bee_and_Honey/) and the Bee Informed Partnership (https://beeinformed.org/programs/management-surveys/) to identify signals in the noise of bee colonies. There is now an equally ambitious study from Austria to add to the mix. Austrians are dogged in registering and tracking their colonies and much else about the environment, and this has led to numerous insights into the impacts of climate, disease, management and stress on Austrian bee colonies. In the most recent study, Sabrina Kuchling and colleagues analyzed data from 129,428 colonies tracked over six years ("Investigating the role of landscape composition on honey bee colony winter mortality: A long-term analysis," (Kuchling et al., 2018)). Colony operations ranged from one or a few hives to 580 hives (one of the country's largest beekeepers). In this study, smaller operations fared better in terms of overwinter losses than larger operations, with a tipping point somewhere above 60 colonies. In addition, the landscape around apiaries played a significant role. One other nugget was that stress factors linked to colony losses were most clear in years that were bad overall. In other words, stresses placed on bees by habitat, climate and disease were additive, eventually pushing more colonies off more ledges. Despite this huge effort, noise persisted in terms of unpredicted anomalies (interactions) including factors that are good predictors of colony loss one year but poor or opposite predictors the next. To this end the authors conclude their extraordinary effort with a humbling, "It also indicates that conclusions drawn from analyses of a single winter should be interpreted with great caution, and further long-term studies are needed to understand honey bee colony losses." Knowing the minds of scientists, and competition for a breakthrough moment, this is not just a plea for more funding but an honest appraisal that, despite a few solid truths (disease=bad, stress=bad, forage=good) we are attempting to manage a social organism that is affected by rules we do not yet fully understand.

Ricigliano, V.A., Mott, B.M., Floyd, A.S., Copeland, D.C., Carroll, M.J. & Anderson, K.E. (2018). Honey bees overwintering in a southern climate: Longitudinal effects of nutrition and queen age on colony-level molecular physiology and performance. Scientific Reports, 8. DOI:10.1038/s41598-018-28732-z.

Corona, M., Velarde, R.A., Remolina, S., Moran-Lauter, A., Wang, Y., Hughes, K.A. et al. (2007). Vitellogenin, juvenile hormone, insulin signaling, and queen honey bee longevity. Proceedings of the National Academy of Sciences of the United States of America, 104, 7128-7133. http://www.pnas.org/content/104/17/7128.

United States Department of Agriculture. National Agriculture Statistics Service. Received at https://www.nass.usda.gov/Surveys/Guide_to_NASS_Surveys/Bee_and_Honey/.

Bee Informed Partnership. Management Surveys. Received at https://beeinformed.org/programs/management-surveys/.

Kuchling, S., Kopacka, I., Kalcher-Sommersguter, E., Schwarz, M., Crailsheim, K. & Brodschneider, R. (2018). Investigating the role of landscape composition on honey bee colony winter mortality: A long-term analysis. Scientific Reports, 8. DOI:10.1038/s41598-018-30891-y.

31

WINTER STIRRINGS

Some people say spring starts with the mating of great-horned owls in late December. These people are greatly outnumbered. For most of us, even February is a month of darkness and cold extremities. Still, it is the first month for wishful thinking and definitely a month for all plans and equipment to be in place for a successful takeoff of new packages and overwintered survivors alike. Your bees are stirring by then as well, and there is new research detailing just how much they are doing to be ready for spring flowers.

In temperate regions, including much of North America, worker honey bees are rarely seen outside during winter. They might search for the few available flowering plants, but mostly they will defecate and return home promptly. This does not mean bees are ignoring the oncoming spring and the need to renew and rebuild. In fact, colonies often start bouts of egg-laying and brood rearing in the middle of winter. It is an interesting management and breeding problem to sort out whether mid-winter brood rearing harms or hurts colonies, whether certain breeds are more prone to flipping the brood switch mid-winter, and the specific cues bees use to start their engines.

Alphonse Avitabile (still mentoring in Connecticut and the co-author of a leading beekeeping book) sacrificed colonies monthly through a Connecticut winter for a 1978 study in the *Journal of Apicultural Research* entitled 'Brood rearing in honey bee colonies from late autumn to early spring' (Avitabile, 2015). Avitabile described substantial winter brood rearing, with sealed brood averages in the thousands per colony from January onward. This was despite the presence of a true winter in his apiaries (average high temperatures of 41, 36, and 38 degrees Fahrenheit for December, January and February, respectively, today and perhaps even colder in the mid-1970's). Similarly, Lloyd Harris followed brood production in Canadian honey bee colonies entering winter in Manitoba, Canada, also describing his findings in the *Journal of Apicultural Research*

(Harris, 2009). Italian bees from California were subjected to mean 'high' temperatures of 18° Fahrenheit by the time the study ended in December. Still, they persisted in egg laying, showing an average of ca. 1000 sealed brood cells when sampled on December 5. Soon after, they were moved to warmer conditions in a climate-controlled warehouse (43 degrees, constantly), and brood numbers expanded and continued until spring (as described in a follow-up paper from 2010 in the same journal (Harris, 2010)).

Fabian Nurnberger and colleagues used an experimental approach to determine when and why bees restart brood rearing in late winter. They describe their results in the open-access journal *PeerJ* in a 2018 article "The influence of temperature and photoperiod on the timing of brood onset in hibernating honey bee colonies" (Nürnberger et al., 2018). These

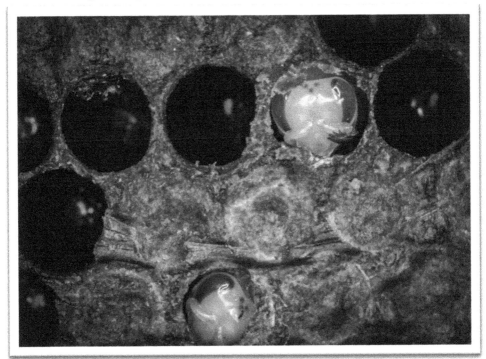

Varroa in bee cell, HFB

researchers followed honey bee colonies in Würzburg, Germany (average high temperatures of 39, 37 and 41°F in December, January and February). They used controlled rooms to manipulate both temperature and the day length perceived by bees. While there were complicated interactions between forced day-length and temperature, they showed in general that temperature was the strongest predictor of the initiation of brood rearing. Once bees committed to brood rearing, they continued to do so even when temperatures were reduced substantially, and the authors propose this as a risk to rebooting brood rearing in the face of an unpredictable climate.

Many beekeepers treat their colonies for *Varroa* mites mid-winter, especially with oxalic acid treatments, which are highly effective against exposed mites but ineffective against mites in sealed cells. If treated colonies harbor patches of sealed brood, oxalic acid treatments could miss substantial numbers of mites. Hasan Al Toufailia and Francis Ratnieks in England addressed this concern. Monitoring colonies in Sussex (high temperatures of 43, 41 and 41°F in December, January and February, respectively) they confirmed that December is the quietest month in terms of brood activity, with between 9 and 52% of colonies having any brood at all across four study years. Variation across years in December brood incidence likely reflects warmth in late fall and continued pollen availability. Substantial brood rearing began in January, where all colonies in each of four years had some sealed brood, averaging 1400 cells each across all years. From a management standpoint, oxalic acid treatments in December would have a more lasting impact on mite levels than treatments in January or any other month.

Another adverse outcome of winter brood rearing is that female mites could both increase their progeny and improve their own health. While some *Varroa* mites no doubt survive months hitchhiking on adult bees, *Varroa* populations as a whole suffer severe winter declines as female mites reach their limits and die. Winter brood provides a significant bridge for declining mite populations. Crudely, if there are 1400 sealed worker brood during January, approximately 110 bees will emerge daily, or 3410 in the whole month. Assuming 25% of these cells contain mites, with 2.5 female mites/cell emerging on average (mom included) these winter brood cells can be factories for >2100 new mites in January and maybe twice as many in February. These mites are younger and presumably more fecund than mites born months earlier. And their moms might benefit as well, since a bout of reproduction involves feeding on plump bee pupae, arguably a richer food resource than overwintering adult bees.

Early starts on brood rearing are likely to be positive on the whole, since a younger and larger bee population will be ready for spring flowers. Still, there is a downside in terms of mite numbers and compromised mite treatments. Next year, start your spring oxalic treatments in early December, before the owls mate.

Avitabile, A. (2015). Brood Rearing in Honeybee Colonies from Late Autumn to Early Spring. Journal of Apicultural Research, 17, 69-73. doi.org/10.1080/00218839.1978.11099905.

Harris, J.L. (2009). Development of honey bee colonies on the Northern Great Plains of North America during confinement to winter quarters. Journal of Apicultural Research, 48, 85-90. https://doi.org/10.3896/IBRA.1.48.2.01.

Harris, J.L. (2010). The effect of re-queening in late July on honey bee colony development on the Northern Great Plains of North America after removal from an indoor winter storage facility. Journal of Apicultural Research, 49, 159-169. https://doi.org/10.3896/IBRA.1.49.2.04.

Nürnberger, F., Härtel, S. & Steffan-Dewenter, I. (2018). The influence of temperature and photoperiod on the timing of brood onset in hibernating honey bee colonies. PeerJ, 2018. DOI: 10.7717/peerj.4801.

32

SPRING GREENING

I am not known for having a green thumb and except for one year of ambitious subsistence gardening (doomed by goats), our limit for many years has been a small herb and tomato plot. I am motivated to attract bees though, and am surrounded by people who know how to do that. The USDA- ARS Bee Research Laboratory has a providential pollinator garden thanks largely to the efforts of Dr. Francisco Posada (garden photos taken by Peggy Greb, USDA-ARS) and it is an excellent refuge for both bees and researchers.

Dr. Posada in the USDA-ARS Bee Research Lab pollinator garden, PG

To better understand all the greenery, I plowed through several papers by colleagues who are testing new ways to keep pollinators safe and well fed. These colleagues are best known for studying the bug side of the pollination equation, but they have teamed up with plant experts for a little, ughh, cross-pollination. So, it is with a recent paper by graduate student Pierre Lau and colleagues at Texas A and M University. In

"Seasonal variation of pollen collected by honey bees (*Apis mellifera*) in developed areas across four regions in the United States," (Lau et al., 2019) the team analyzed hive-collected pollen grains in California, Michigan, Texas, and Florida. Their focus was the realm of gardeners and city planners, areas with only 20-35% undeveloped land. Using pollen traps, they identified the first 200 pollen grains per sample to species or, in some cases, the next level up. Suburban California proved to be most diverse, followed by Florida, then Michigan and Texas. Overall, bees collected more diverse forage in the spring, and an even mix of pollen from woody plants and trees versus herbs. Colonies were more faithful than expected to particular species on a weekly timescale, suggesting fidelity by individual bees or recruiters, or perhaps subtle preferences at the colony level related to immediate needs. If you are planning a garden, or shopping among the numerous pollinator seed mixes for your area, the authors of this open-access article provide helpful information for four divergent habitats.

Going straight to the city, Dr. David Lowenstein and colleagues mapped out over 1000 pollinator visits to flowers in dozens of Chicago parks and gardens (Lowenstein et al., 2018) After confirming the attractiveness of well-known weeds and ornamentals, especially perennial plants, they highlight the profound lack of interest by pollinators for three plant groups (petunias, impatiens, and marigolds) loved by gardeners for attractiveness and hardiness. Similarly, noted bee biologist Dr. Dave Goulsen paired up with botanist Ms. Rosi Rollins (http://www.rosybee.com/research) to quantify the attractiveness of 111 putative bee magnets (Rollings, 2019). They confirmed that many plants on 'pollinator mix' hit lists were indeed attractive to honey bees, bumble bees and a diverse set of additional pollinators, but also recognized plants such as calamint, a plant that was the top choice among many bees but not among gardeners or list-makers. Perennial geraniums were super attractive to bees as well, but these are distinct from their shinier annual relatives, which tend not to seduce bees. Several other plant groups showed one species with high attractiveness and close relatives which were not valued by the bees. In addition, the authors point out that plants found on other bee-friendly lists in fact received few or no visits. The study was carried out in the United Kingdom but the open-access paper should be relevant for bee gardens in the US as well. In fact, >75% of their vetted plants are not UK natives, similarly to ornamental plants elsewhere.

At the farmscape level, Adam Dolezal and colleagues at the University of Illinois and Iowa State University carried out a two-year study showing how bees deal with the ups and downs of intensive agriculture (Dolezal et al., 2019b). Colonies were established adjacent to 20 soybean farms in total, across two years. Ten of these farms were surrounded by croplands (with 84% of the bee-flight space comprised of corn or soy, not unusual in Iowa) while ten were surrounded by other land uses, with around 40% of the acreage in corn and soy.

Soy fields with prairie strips, OKM

Colonies were established some weeks after soy planting to minimize the effects of chemically coated seeds. Colonies in the ag-intensive plots did best, reaching around 25 frames of adult bees and about 25 kilograms of hive weight by mid-August, versus 15 frames and 15 kilograms in the low-cropping plots. However, by late August, colonies at both sites collapsed significantly, reaching identically low weights by the end of the season. In all plots, bees collected nectar, but rarely pollen, from soy plants and relied on pollen from clover and other neighboring legumes. When these surrounding flowers went into steep decline, colonies began to shrink. In one year, the researchers kept colonies in an ag setting (their own land-grant University farm) until August and then sent them to pasture in a diverse prairie setting. The bees whooped it up there, increasing their hive weights and health substantially compared to the colonies left in place on the farm. The authors argue that these moves, if timed perfectly, can give beekeepers (and soy growers) the benefits of soy pollination through much of the summer and a late-season boost after that. The 'STRIPS' project in Iowa (same affiliation as the Feast-Famine folks, https://www.nrem.iastate.edu/research/STRIPS/content/faq-how-can-i-get-prairie-strips-my-farm) favors adding these Prairie strips directly to the farmscape in the hopes that bees can take advantage of an extended and more diverse bloom without being moved.

While changes to backyards, parks, and farmscapes are important, there are strong efforts to increase forage opportunities at even larger scales. Policy and planning groups like the Pollinator Partnership (https://www.pollinator.org/) and the Bee and Butterfly Habitat Fund (BBHF; https://beeandbutterflyfund.org/about-us) are focused on developing planting guides for diverse habitats and increasing acres of bee forage on public and private lands, respectively. In particular, the BBHF has had great success in improving pollinator resources on large-scale farms and pastures. In the science realm, the Dr. Clint Otto and colleagues at the USGS have carried out numerous studies showing the value of rangelands, and seeding with flowering plants, for bees in the Dakotas (I reviewed one such effort in 2017 in essay 11, above). Dr. Vincent Ricigliano and USDA colleagues also showed the benefits to honey bees of improved rangeland habitats (Ricigliano et al., 2019), and these benefits no doubt apply to other important pollinator species as well.

Like many environmental efforts, improving bee forage can be tackled at various scales, from a well-timed container garden of flowering plants, or a water source for bees, to policy that impacts thousands of acres of potential bee space. Many more like-minded groups are working to increase safe bee forage and these efforts are taking root, improving the chances for honey bees as well as other pollinators.

Lau P, Bryant V, Ellis JD, Huang ZY, Sullivan J, Schmehl DR, Cabrera AR & Rangel J (2019) Seasonal variation of pollen collected by honey bees (Apis mellifera) in developed areas across four regions in the United States. PLoS ONE 14. doi:https://doi.org/10.1371/journal.pone.0217294.

Lowenstein DM, Matteson KC & Minor ES (2018) Evaluating the dependence of urban pollinators on ornamental, non-native, and 'weedy' floral resources. Urban Ecosystems 22: 293-302. doi:https://doi.org/10.1007/s11252-018-0817-z.

Rosybee Plants for bees. Received at (http://www.rosybee.com/research.

Rollings R (2019) Quantifying the attractiveness of garden flowers for pollinators. Journal of Insect Conservation 23: 803-817. doi:https://doi.org/10.1007/s10841-019-00177-3.

Dolezal AG, St. Clair AL, Zhang G, Toth AL & O'Neal ME (2019) Native habitat mitigates feast–famine conditions faced by honey bees in an agricultural landscape. PNAS 116: 25147-25155. doi:DOI: 10.1073/pnas.1912801116.

Iowa State University Research. Received at https://www.nrem.iastate.edu/research/STRIPS/content/faq-how-can-i-get-prairie-strips-my-farm.

Pollinator Partnership. Received at https://www.pollinator.org/.

The Bee & Butterfly Habitat Fund. Received at https://beeandbutterflyfund.org/about-us.

Ricigliano VA, Mott BM, Maes PW, Floyd AS, Fitz W, Copeland DC, Meikle WG & Anderson KE (2019) Honey bee colony performance and health are enhanced by apiary proximity to US Conservation Reserve Program (CRP) lands. Scientific Reports 9. doi:https://www.nature.com/articles/s41598-019-41281-3.

CLOSING

33

#BEEOPTOMISM: WHAT IF THE HONEY SUPER REALLY *IS* HALF FULL?

My ears perked up recently on hearing several leaders in environmental science talking, for once, about some silver linings in our environment. This 'Earthoptimism' ethic (e.g., https://earthoptimism.si.edu/) does not deny the huge impacts, many negative, our species is having on the planet. It doesn't even claim that things can't get worse. Instead, Earthoptimism strives to point out that much remains. More importantly, the movement seeks foot soldiers to preserve what remains by putting all their energy into solutions. Many of us only respond to recruitment fliers when the outcome has some hope of being positive. Nancy Knowlton, an ocean

Pollinating honey bee, HFB

researcher who is driving the Earth Optimism movement, has every reason to be pessimistic. She is well aware of coral bleaching, changing climates, and those swirling masses of ocean plastic that show up on YouTube (https://www.youtube.com/watch?v=fDHuPjx0aPQ). Despite these challenges, she is promoting the more powerful view that there is much to be done and we are the ones to do it. As she says, "Big problems without solutions leads to apathy. Big problems with solutions leads to action."

The bee world deserves and, more importantly, needs a bit of optimism to start the new year. This doesn't let bad beekeeping, bad stewardship, or even (gasp) bad research off the hook, but points to what is working and especially to all that is worth saving; it is absolutely true that we need healthy populations of pollinators for our food supply and environment. Achieving this requires improving the tools used by beekeepers (for honey bees) and habitat stewards (for honey bees as well as the remaining bees) to maintain these populations. These tools range from science to management and regulation.

I have always been most comfortable in the optimists' club, despite occasional bad luck and outright failures by myself, and those around me. We all have formative experiences, and I spent a year of my early teens in and out of hospitals with pretty severe kidney disease. It was apparently some sort of puzzle for the day's medicine, or I wasn't responding, or both, but I spent months and months as just another sad kid with a catheter. We Evans' are pretty stoic and I don't remember my caring parents ever explaining why I was still sick, and I assume they were just as scared as I was. Enter Dr. Liliedahl, a pediatrician who not only had a plan of attack but an overwhelmingly infectious optimism. I distinctly remember my parents looking puzzled and perhaps thinking, "what is this victory of which he speaks?" But his spirit dragged us along for additional months and procedures and eventually things worked out fine.

So…what can we be optimistic about in the world of honey bees and other pollinators?

Here are five things, in no particular order:

1)   We have a great corps of passionate new beekeepers on the scene, and they're good at communicating in real time. They are using Twitter,

Instagram, and other avenues to communicate and improve their beekeeping.

2)   Ditto that for new researchers. I am in awe at how quickly and effectively my younger colleagues transmit their latest thoughts and results, leading to some real and rapid advances, and decreasing redundant studies. This #oldguy is impressed.

3)   We have new methods, treatments, and diagnostics that are being used to peel back the darkness, from super-sensitive assays of chemical levels in bees and hive products (e.g., (Mitchell et al., 2017)) to genetic screens for stress and disease (Schurr et al., 2017), to hive monitors like the one in my backyard that reports to the web day and night (thanks www.hivetool.net).

4)   We have systematic surveys that are amassing great insights into what happens just before colonies thrive or die, where those colonies are positioned, and how their beekeepers treated them. In the U.S., the Bee Informed Partnership and affiliated efforts (www.beeinformed.org) and the USDA-NASS national bee health survey (www.nass.usda.gov) provide fascinating and robust insights into losses from prior years, and they are only getting better. Worldwide, the COLOSS network (www.coloss.org) helps coordinate and advertise similar efforts and fights to make survey datasets available to researchers and the public.

5)   Most importantly, we work with a truly resilient organism that needs help right now, but can also do a lot on her own.

I am afraid that ending with this essay will raise hackles among those who see the many challenges to honey bees and other pollinators. I do not pretend we live in in perfect times. The threats to pollinators are numerous. But to those who say tone down the smiles until things get better, I say fiddlesticks, give me Pharrell any time over Morrissey. Can't wait until spring, and counting on better times for beekeepers.

Smithsonian Conservation Commons. Earth Optimism Summit. Received at https://earthoptimism.si.edu/.

"Can this project clean up millions of tons of ocean plastic?," YouTube video, 00:08:26, Posted by "PBS NewsHour," Aug 14, 2016, https://www.youtube.com/watch?v=fDHuPjx0aPQ.

Mitchell, E.A.D., Mulhauser, B., Mulot, M., Mutabazi, A., Glauser, G. & Aebi, A. (2017). A worldwide survey of neonicotinoids in honey. Science, 358, 109-111.

Schurr, F., Cougoule, N., Rivière, M.P., Ribière-Chabert, M., Achour, H., Ádám, D. et al. (2017). Trueness and precision of the real-time RT-PCR method for quantifying the chronic bee paralysis virus genome in bee homogenates evaluated by a comparative interlaboratory study. Journal of Virological Methods, 248, 217-225.https://doi.org/10.1016/j.jviromet.2017.07.013.

Real Time Honeybee Monitoring tool. Received at www.hivetool.net.

The Bee Informed Partnership. Received at www.beeinformed.org.

United States Department of Agriculture. National Agriculture Statistical Service. Received at www.nass.usda.gov.

COLOSS Honeybee Research Association. Received at www.coloss.org.

34

#BEEOPTOMISM 2.0: LET'S NOT GO VIRAL

I am a huge fan of the #Beeoptimism movement. Built on a similar effort to see hope for our Earth in spite of it all (https://earthoptimism.si.edu/), Beeoptimism focuses on the 50% of colonies that survive annually, the honey jars and supers half full, and the hope that someone (I am guessing someone younger and smarter than me) will piece together real solutions for bee health. Below are five reasons to be optimistic. These come from individuals and teams who have helped make a little more sense of bee life, or have come up with science-based solutions that seem ready to take up.

1) I continue to marvel at the new insights coming from colony-level ways to monitor bee foraging, life, and death. For ages, we have known

Returning foragers, JDE

how to help or hurt bees in the lab, with acute stresses or disease agents. We have also known how to monitor colonies for diseases, growth, and decline. Only now are many labs conducting high-quality field trials by tagging hundreds of individual bees with Radio-frequency identification, or RFID, tags (think low-tech ankle bracelets) to show when bees do things right or wrong over their lifetimes. Coupled with accurate scales to weigh individual bees after they land from trips, cameras to check their appearances, and analytical tools to chew through all the data, this advance will help address a number of challenges to healthy bee hives.

2) The Bee Informed Partnership (www.beeinformed.org) and the USDA-NASS National Honey Bee Survey (www.nass.usda.gov) make the list for collecting survey information (thanks, beekeepers!) and data from numerous other sources as a means of monitoring the industry and what is working for bees. They have been joined by 'Beescape,' an ambitious effort from Pennsylvania State University and several partners (www.beescape.org). Beescape focusses on neighborhoods around apiaries, or potential apiaries. Currently Beescape allows people to obtain information on land use, seasonal forage resources, wild bee nesting habitat, and pesticide risk around their apiaries. The results could be quite practical if you are a beekeeper choosing between a backyard apiary or hitting up an uncle 30 miles away. The Beescape team is now modeling how these factors, and weather conditions, influence honey bee colony survival and performance, and plan to release a winter survival prediction tool in early 2020. Beescape also offers interactive programs for volunteers interested in sharing data on their honey bee colonies to help improve these models.

3) It is hard not to be optimistic about the range of natural products being vetted right now to improve bee health. Our group has jumped on this topic, and even wrote a recent manifesto and recipe book for researchers (https://www.mdpi.com/2075-4450/10/10/356), but the field of scientists looking into this is diverse, broad, and not necessarily new to the hunt. Groups seeking ways to improve bee health through natural products include plant forage experts, longtime researchers in bumble bee ecology, experts in mushrooms, and chemists who are disentangling the many components of propolis. Chemistry is chemistry, and some sure-fire natural products are likely to be *bad* for honey bees, while nearly all of the rest will have no real impact on bee health. Still, it is impossible not to be optimistic that somewhere out there is a perfect

extract, or molecule, that makes bees resilient in the face of viruses and other diseases. It is also entirely possible that, as with propolis, bees have recognized the benefits of specific natural products for millions of years. In that case, our job is simpler in that we just have to confirm what they know and shorten the trip to get these goodies into the hive.

4)   Honey adulteration remains a big deal for honest beekeepers and consumers. In the past year, government agencies in the US and elsewhere have tested new technologies to identify funny honey and hold people to account for it. One recent study, by Huijun Wang and colleagues in China and the United Kingdom, accurately identified different honeys to their plant sources, and spotted adulterants such as corn and rice syrup at the level of 5% ("A novel methodology for real-time identification of the

Clover-enriched pasture, JDE

botanical origins and adulteration of honey by rapid evaporative ionization mass spectrometry," (Wang et al., 2019)). In "Use of NMR applications to tackle future food fraud issues" (Sobolev et al., 2019), Anatoly Sobolev and colleagues discuss ways of economizing the machines and diagnoses needed to validate honey and other foods. While these technologies are part of an arms race with those hoping to evade them, the science seems to be winning for now.

5) I will end with a story for the future that has sucked me in despite being way out of my scientific wheelhouse. First, it is undeniable that bees of all sorts do better with fields of flowers than fields of turf. For bee fans, the 50+ million acres of turf in the US (https://agamerica.com/turfgrass-industries/) represent a blank canvas. Turf ranks among the top three US 'crops,' with soybeans and corn. As with soy and corn, tapping into the turf environment by making it more bee friendly would be a really big deal. Many property owners seem quite satisfied with a turf lawn but others, given a nudge, would rather turn their lawns into a more functional space by supporting bees and other wildlife. So-called 'bee lawns,' with a little effort, can be beautiful and enriching for honey bees and other beneficial insects (https://www.beelab.umn.edu/learn-more/beelawn). Most lawns, with small changes in mowing and herbicide behavior, will support white clover and other flowers for much of the year. Mr. Steve Hess, of Indiana, runs his own environmentally focused pest control business. He is an entomologist who walks the fine line between removing pest insects and protecting his beloved bees. He was acutely aware that his high-speed commercial lawnmower was mulching numerous bees, and there was no way that either he or his bees could avoid collisions. Knowing the turf stats above, he also calculated the country-wide bee losses due to lawn maintenance and was appalled. Mr. Hess' conscience and creativity have led to one possible solution. He has invented a cattle guard of sorts for riding mowers, meant to gently lift bees above the mower deck just before they are sucked into the blades. He and his engineering partners continue to improve the details, but a patented model that we have both tested seems durable and likely to save some bee lives. The testing protocol consists of strapping a video camera in front of the deck and mowing fields of clover with and without the guides. Despite many acres of footage, and one mulched iPhone, it is surprisingly hard to quantify how many bees take flight versus drop to the turf and under a mower deck. Still, we have seen enough of the latter, and enough improvement with the guides, to indicate that this will be a great holiday present for landscapers and bee lovers someday. Mr. Hess remains passionate about this and is one of the hardest working people I have met. If you would like to learn more about these 'Bee Guards', you can contact Mr. Hess directly at idbugu7@gmail.com.

So.. go out there with optimism and realism and double your honey this year, or at least plant some flowers and mind the bees.

Smithsonian Conservation Commons. Earth Optimism Summit. Received at https://earthoptimism.si.edu/.

Bee Informed Partnership. National Management Survey. Received at https://bip2.beeinformed.org/survey/.

United States Department of Agriculture. National Agriculture Statistics Service. Received at https://www.nass.usda.gov/Surveys/Guide_to_NASS_Surveys/Bee_and_Honey/.

Beescape. Get a bee's eye view of your landscape. Received at http://www.beescape.org).

Tauber , J. P., et al. (2019). "Natural Product Medicines for Honey Bees: Perspective and Protocols." Insects 10(10): 356.

Wang, H., et al. (2019). "A novel methodology for real-time identification of the botanical origins and adulteration of honey by rapid evaporative ionization mass spectrometry." Food Control 106.

Sobolev, A. P., et al. (2019). "Use of NMR applications to tackle future food fraud issues." Trends in Food Science & Technology 91: 347-353.

The Importance of the Turfgrass and Turf Managemment Industries. Received at https://agamerica.com/turfgrass-industries/.

University of Minnesota. Bee Lawns. Received at https://www.beelab.umn.edu/learn-more/beelawn.

PHOTO CREDITS

AB, Image by Andrew B. Barron, Macquarie University

BB, Image by Belén Branchiccela, Instituto de Investigaciones Biológicas Clemente Estable

GB, Photo by Gary Bauchan, USDA Agricultural Research Service

HFB, Photo by Humberto F. Boncristiani, www.insidethehive.tv

RB, Photo by Robert Brodschneider, University of Graz

JE, Jim Erlund, USDA Agricultural Research Service

JDE, Photo by Jay D. Evans, USDA Agricultural Research Service

PG, Photo by Pamela Gregory, USDA Agricultural Research Service

YH, Photo by Yanbo Huang, USDA Agricultural Research Service

DL, Photo by Dawn L. Lopez, USDA Agricultural Research Service

OKM, Photo by Omar de-Kok Mercado

Jay Evans studied ants at Princeton University and the University of Utah before developing a passion for honeybees. He is currently Research Leader for the United States Department of Agriculture Bee Research Laboratory in Beltsville, MD. The BRL is focused on the development of management strategies to help honey bees thrive in the face of disease, chemical stress, and inadequate forage. Current projects involve honey bee immunity, interactions among stress factors, and the development of novel, safe, controls for mites and pathogens. This book was born from a series of love poems for scientists and beekeepers.

CPSIA information can be obtained
at www.ICGtesting.com
Printed in the USA
LVHW071449010920
664728LV00006B/232